T0234506

SOCIAL MEDIA AND THE CONTEMPORARY CITY

The widespread adoption of smartphones has led to an explosion of mobile social media data, more than a billion messages per day that continuously track location, content, and time. *Social Media in the Contemporary City* focuses on the effects of social media on local communities and urban space in a variety of political and economic settings related to social activism, informal economic activity, public art, and global extremism.

The book covers events ranging from Banksy art installations, mobile food trucks, and underground restaurants, to a Black Lives Matter protest, the Christchurch mosque shootings, and the Pulse nightclub shooting. The interplay between urban space, local community, and social media in each case study requires diverse methodologies that are both computational (i.e. machine learning, social network analysis, and natural language processing) and ethnographic (i.e. semi-structured interviews, thematic analysis, and site analysis). The book views social media not as a replacement for the local community or urban space but rather as a translation of the uses and meanings of all three realms.

The book will be of interest to students, researchers, and instructors in a number of disciplines including urban design/planning, media studies, geography, and communications.

Eric Sauda is a professor of architecture and an adjunct faculty at the Charlotte Vis Center and the School of Data Science UNC Charlotte. His research interests include interactive environments, digitally augmented performance, and social media in architecture and urban settings. Professor Sauda's work has been published in the *Journal of Architecture Education, Journal of American Planning Association, New Media & Society, Social Media + Society, The Routledge Companion to Critical Approaches to Contemporary Architecture,* and *Planning Support Science for Smarter Urban Futures.*

Ginette Wessel is an assistant professor of architecture at Roger Williams University. Her primary research interests include contemporary issues of urban development, with an emphasis on social equity, sustainability, and communication technology. Dr. Wessel's research has been published in *The MIT Press, Journal of American Planning Association, New Media & Society, Journal of Urban Design,* and *Participatory Urbanisms.*

Alireza Karduni is a postdoctoral scholar at Northwestern University's Department of Computer Science. His research connects human computer interaction and computational social sciences. He studies how people interact with social media data under uncertainty. Dr. Karduni's work has been published across multiple disciplines, in venues such as *Transactions in Visualizations and Computer Graphics, Journal of American Planning Association,* and *Social Media + Society.*

SOCIAL MEDIA AND THE CONTEMPORARY CITY

Eric Sauda, Ginette Wessel,
and Alireza Karduni

Routledge
Taylor & Francis Group

NEW YORK AND LONDON

First published 2022
by Routledge
605 Third Avenue, New York, NY 10158

and by Routledge
2 Park Square, Milton Park, Abingdon, Oxon, OX14 4RN

Routledge is an imprint of the Taylor & Francis Group, an informa business

Library of Congress Cataloging-in-Publication Data
Names: Sauda, Eric, author. | Wessel, Ginette, author. |
Karduni, Alireza, author.
Title: Social media and the contemporary city / Eric Sauda,
Ginette Wessel, and Alireza Karduni.
Description: Abingdon, Oxon ; New York, NY : Routledge, 2022. |
Includes bibliographical references and index. |
Identifiers: LCCN 2021025449 (print) | LCCN 2021025450 (ebook) |
ISBN 9780367459109 (hardback) | ISBN 9780367902506 (paperback) |
ISBN 9781003026068 (ebook)
Subjects: LCSH: Urbanization—Social aspects. | Information
technology—Social aspects. | Social media—Political aspects. |
Sociology, Urban.
Classification: LCC HT153 .S274 2022 (print) | LCC HT153 (ebook) |
DDC 302.23/1—dc23
LC record available at https://lccn.loc.gov/2021025449
LC ebook record available at https://lccn.loc.gov/2021025450

ISBN: 978-0-367-45910-9 (hbk)
ISBN: 978-0-367-90250-6 (pbk)
ISBN: 978-1-003-02606-8 (ebk)

DOI: 10.4324/9781003026068

Typeset in Times New Roman
by codeMantra

CONTENTS

ILLUSTRATIONS

ACKNOWLEDGMENTS

This book is the result of a long collaboration between the authors, each bringing a different perspective to the work. It began with a curiosity about how the contemporary city is influenced by the spread of information and communication technologies. This book focused in particular on how the spread of mobile computing prompted by the rapid and wide proliferation of smartphones may have accelerated and shifted this influence. Anyone familiar with these trends would suspect something might be afoot.

The seduction of new technology can be difficult to resist, even if one suspects its impact may be overblown by its advocates. New technology has arrived before and while it changes things, it never changes *everything*. But the only thing more problematic than messianic optimism is hidebound conservatism, the insistence that all important things are just as they were 10 years ago, 100 years ago, 1000 years ago.

The goal of this book then is to see how things change and how contemporary culture adapts to new circumstances when faced with what McLuhan described as a new extension of man.

Much of this work has been advanced through a series of graduate seminar courses in the Design Computation Dual Master's degree program in Architecture and Computer Science at UNC Charlotte. The students brought enthusiasm and energy to these courses and enabled the testing of strategies and methods for the incorporation of data analysis in the study of spatial settings. The authors wish to thank Anna Anklin, Chelsea Hansen, Kacie Ward, Richard Crouch, Chi Zhang, Xunxun Zhang, Janette Billings, Josn Kieb, Marc Remesi, Heather Tarney, Phillip Broszkiewics, Ashley Damiano, Bradford Guillet, Alex Shuey, Benjamin Sullivan, and Brendon Bryant. Graduate research assistantships from the Graduate School at UNC Charlotte provided us with strong support to advance our work. They were vital to our effort and deserve much credit for their contribution. Trevor Hess and Saquib Sarwar contributed early in the research.

During the final research necessary for the manuscript, our research assistants Atefeh Mahdavi Goloujeh and Sanaz Ahmadzadeh Siyahrood worked tirelessly on the difficult and sometimes tedious tasks that allowed the discovery of connection within the data.

The authors also thank their colleagues in both architecture and computer science for the many discussions and critiques of the ongoing work. It is one of the greatest joys of working in a university. Dr. William Ribarsky provided the widest perspective of visualization and data analysis and was an inspiration for our work. Dr. Wenwen Dou, Dr. Remco Chang, Dr. Caroline Ziemkiewicz, Dr.Isaac Cho, Dr. Charles Davis, Dr. Mary Lou Maher, and Dr. John Gero were able to give us vital feedback at different stages of our research.

Early versions of two chapters of this book were previously published and the authors wish to acknowledge these journals. An early version of the Black Lives Matter chapter was published as "Anatomy of a protest: spatial information, social media, and urban space" in *Social Media+ Society 6*(1). An early version of the mobile food truck chapter was published as "Revaluating urban space through tweets: An analysis of Twitter-based mobile food vendors and online communication" in *New Media & Society, 18*(8). Some work on an early Banksy installation used in the book was based on research from a graduate seminar by Brendon Bryant.

This book would not have been possible without the support of Dean Doug Hague and the School of Data Science at UNC Charlotte. They responded to the requests for more and more data with generosity and attention.

The authors wish to especially thank Neda Kardooni for her willingness to co-author the chapter on Iranian vendors. Her deep understanding, careful study of examples in Tehran, and scholarly background in anthropology were vital to this chapter that provided an expanded cultural perspective to the book.

Dr. Alireza Karduni wishes to thank his parents, Vahideh Gandomikal and Habibollah Kardooni, for their patience, especially during the days of the work on this book, before seeing their son after eight years. He would also like to thank his siblings Hoda, Neda, and Roozbeh Kardooni who supported him throughout this work. He also would like to thank Dr. Wenwen Dou, his PhD adviser, for her patience and flexibility during the final days of his dissertation as he was simultaneously working on this book and his dissertation. He wishes to sincerely thank his friends, Alireza Bahramirad, and Alexandra Mcnally, who graciously provided him a comfortable room to stay in and took care of him while he was recovering from Covid. Finally, he would like to thank his friends and family in Iran and other parts of the world, who through many hours of voice chats on Discord and online games, helped him get through the loneliness of immigration.

Dr. Ginette Wessel wishes to thank Ken Lambla, the founding Dean of the College of Arts & Architecture at UNC Charlotte, for his continued support and gratitude for this research over many years. She would also like to thank Margaret Crawford, Director of Urban Design and Professor of Architecture at UC Berkeley, whose passion for the changing dynamics of urban space has influenced her scholarship with new pathways of research. She also wishes to thank her colleagues at Roger Williams University for their valuable time and feedback on drafts and presentations of this work. In particular, thanks to Stephen White for encouraging me to purse collaborative research. Finally, this work would not be possible without the loving support of her husband Nicholas and daughter Sydney.

Eric Sauda wishes to thank Chris Jarrett, the former Director of the School of Architecture at UNC Charlotte, for his help in arranging on very short notice a reassignment of duty that provided time that was invaluable to the research and writing of this book. He also wants to express deep appreciation to his coauthors. It is great fun to work with people who are unafraid to challenge each other and finish with a final result that is greater than anything they could have done alone. And he most especially wants to thank his wife Georgette who in many ways great and small supported him during his entire career and especially during the hard labor of completing this book.

1

INTRODUCTION

ILLUSTRATION 1.1 Social media in the city.

The rise of smartphones, beginning in 2006, altered the way humans engage with and encounter cities. Today, smartphones are essential components of our daily apparel (see Illustration 1.1) that collect information about our location, our activities, and our time. Social media applications gather much of this information. In 2013, nearly 500 million messages were sent on Twitter and 80% of them on mobile devices.[1] Similar usage is true for WhatsApp, Facebook, Instagram, and other mobile social media platforms.

DOI: 10.4324/9781003026068-1

While social media predates smartphones, its use and ubiquity have greatly expanded. The simultaneous growth of mobile smartphones and social media applications has led to a rich new source of data about the contemporary city.

Although there is widespread interest in the effects of information and communication technology on the city (often filled with either enthusiasm or apprehension), there is no detailed framework for understanding how technology may alter the social and spatial structure of the cities and the meaning and use of urban space. Such a framework would need to include both a theory to guide methodology and a means of processing the vast amounts of empirical data inevitable to conduct such a study. A framework would also need to situate itself within the range of existing, normative methodologies as either an alternative or a complement.

Mobile social media has already made a considerable impact on our everyday lives, transforming the way events occur in cities. Urban events in this discourse are specific *activities* that take place in *particular types of urban spaces* at specific *times*. The concept of *events* formalizes the way mobile social media alters our relationships with urban space by affecting the types of activities that take place and the temporal rhythm of when urban space is occupied. An event requires three overlapping methods: a spatial analysis of urban development, an ethnographic study of the meaning of social space, and an analysis of the flow of mobile social media data in the city. This book explores examples of such changes through advanced methods of data analysis, ethnographic research, and the development of a theoretical framework.

Mobile social media, for example, has become an instrumental part of urban protests. The rapid rate of information propagation allowed activists to plan protests during the Arab spring, "occupy" movements located in cities across the world and protest in other countries including Iran, Spain, and Ireland. Many of these protests were planned and organized in prominent symbolic spaces, but they occurred more spontaneously due to the combination of communication on social media, as well as participation by local activists. There are also many examples of spontaneous protests happening in urban spaces without significant planning.

England-based street artist Banksy has been disrupting urban spaces since the 1990s by creating his art in unknown, marginal urban spaces with an announcement on social media. In 2018, his work gained attention, with a single Instagram post speculating about his stencil art *Season's Greetings* mural on a garage in Port Talbot, Wales. In less than two days, the work was confirmed as Banksy's, and interested individuals from around the world started an online discussion. The unfamiliar location instantly became globally popular to visit. Mobile social media has increased the impact and audiences of Banksy's work. In this case, mobile social media has transformed the garage into a digitally-mediated social space, placing an uncommon activity in an unfamiliar space, that is documented and distributed over social media. The meaning of this urban space is changed from marginal to significantly critical within the art community.

The authors share expertise and research within the fields of architecture, urban design, and planning with additional interest and expertise in computer science, geography, and media studies. This book is intended for all these disciplines.

The introduction frames two important parallel discourses that are relevant to the recent growth of digitally driven activities in cities, albeit in very different ways. The first, *digital urbanism*, includes urban space that is increasingly experienced, accessed, and represented via information technology. Digital urbanism includes an array of analyses, from virtual environments such as chat rooms and smart wayfinding to urban visualization of real-time cell phone data. Opposition to digital urbanism denies that anything important has changed in the contemporary city, believing that the role of digital media is secondary to the creation of urban space. Digital media scholars concerned with cities debate the significance of space in the future but often stray from social questions or concerns.

A second discourse that frames this book includes *social aspects of the built environment* and is based on ethnography to uncover the meaning and motivations for using urban space from the bottom up. Opposed to deterministic views that either neglect or seek to predict social life, bottom-up urbanism (also related to tactical, opportunistic, everyday, temporary, or spontaneous urbanism) offers a critical lens to investigate new forms of urban life that are experimental, spontaneous, or unpredictable, and driven by citizens opposed to special interests. This social viewpoint of urban life highlights the critical role of citizens as autonomous agents capable of reinventing new forms of social urban life.

Digital Urbanism

The arrival of widespread digital media and engagement is true in the most intimate sense. At the level of the body, Hans Moravic[2] has argued that the human body is an outdated system. After comparing robots and humans, he proposes the replacement of the biological by the electronic and the migration of thought into "...the immensities of cyberspace...teeming with unhuman superminds" that he looks forward to without flinching.

Within architecture and urban design, the proliferation of new digital technologies has been an active area of theory and research for at least the last quarter-century. The trilogy of books by William Mitchell (*City of Bits, eTopia, Me++*)[3] marks a convenient beginning to the contemporary investigation of these issues at scales ranging from personal to urban. He announces the death of the city, claiming that the "...city – as understood by urban theorists from Plato and Aristotle to Lewis Mumford and Jane

Jacobs – can no longer hang together and function as it could in earlier times...long live the new, network-mediated metropolis...”[4] Mitchell's writing pulls together a wide range of sources and insights, sketching at a very high level the distinctions he sees between traditional spatially defined urbanism and an emerging vision of digitally mediated environments. These distinctions define a general area of investigation but contain very little about methods that might bridge the divide. He is left to suggest examples of neighborhoods in the city that provide extreme high-speed access for digital devices.

The Smart Cities[5] movement emerged as a loosely defined attempt to integrate new forms of digital information and control into the city. "Smart city" covers a wide range of definitions, but three foci have emerged: smart technology, smart places, and smart systems. Smart technology focuses on the proliferation of sensing technologies that have emerged in the last ten years; this approach is often promoted by hardware manufacturers such as Cisco. Combined with analytic software, it typically aims to solve specific problems such as traffic congestion or water quality through the use of sensors specifically tailored to the problem. Smart places are described as neighborhoods within a larger city that through network access, advanced research facilities, and human capital are likely to lead to advanced thinking and collaboration. The prototypical smart place is the tech square near MIT, but there are now examples throughout the United States and the world. Smart systems refer to cities that are designed or redesigned from the beginning as coordinated across a wide range of digitally controlled systems. Examples include Songdo (South Korea) and PlanIT Valley (Portugal).

Whatever the focus, smart cities are suffused by a faith in top-down systems, assuming a defined set of problems and a carefully orchestrated set of solutions. There are certainly advantages to be gained within specific settings by this approach (traffic may run more smoothly, stream water quality can be monitored continuously, etc.). But the spread of "smart" in the city is just as often more diffuse and less inherently focused than this framework imagines. For example, the identification of a crisis in the city is discovered quicker by Twitter than by custom-built systems.[6] Bottom-up systems such as social media can reveal activities within the city despite their heterogeneous and unstructured content.

Carlo Ratti's *The City of Tomorrow*[7] broadly surveys and synthesizes contemporary approaches to technologically influenced urban design. He proposes a design science approach with a set of proposals for how emerging technologies might be imagined to optimize urban development. To describe this approach, he coined the term futurecraft. This is an updated appeal to Alan Kay's maxim at the Palo Alto Research Center, "The best way to predict the future is to invent it". The strength of *The City of Tomorrow* is its focus on disruptive technologies, suggesting ways in which normative urban understanding and objects might be extended.

Social Urban Space

The absence of social concerns in digital theory scholarship demands a framework that integrates the ways that people and politics shape urban space. While digital theorists prioritize current and future transformations of urban systems, ethnographers and urban sociologists conceptualize space created through a series of actions by groups of people, often at the margins of power. Bottom-up processes that illustrate human agency and allow for unplanned and unregulated activities to occur in urban space are prioritized, opposed to top-down strategies such as regulatory zoning measures or large-scale neoliberal development that situates human activity as subordinate. Urban space in this context changes as the needs and abilities of local residents or participants change over time.

Ideas about the production of social space can be traced to the work of the French Marxist philosopher and sociologist Henri Lefebvre in 1947 with *Critique of Everyday Life – Volume 1: Introduction*[8] and culminating with *The Production of Space* in 1974.[9] The core premise of his work suggests that space gains its meaning not from abstract Cartesian quantities (what Lefebvre calls absolute space) but from its reproduction of the social order of the inhabitants (what he calls social space). There is considerable variation in the use of the terms, "place" and "space"; Lefebvre uses the term social space in the manner that others (particularly those from anthropology) use "place".

Lefebvre's article "The Right to the City", originally published in *Le Droit á la ville* in 1968, expressed his deep concerns with the expansion of capitalism and the decline of the *oeuvre*.[10] *Oeuvre*, a term suggestive of a performance, refers to the "information, symbolism, the imagery and play" in daily life (63:147). As such, "the right to the *oeuvre*, to participation and appropriation, are implied in the right to the city" (63:174). In this work, Lefebvre demands a renewed right to urban life where "the working class can become the agent, the social carrier or support of this realization" (63:158). Lefebvre's concern for the erosion of social life in the city due to the powerful forces of capitalism marked a critical realization that agency is unequal.

In 1974, Lefebvre's book *The Production of Space* became well known for its careful analysis of social space. At a basic level, Lefebvre recognized that modes of thinking about space are physical, mental, and social and that they should be seen simultaneously as real and imagined, concrete and abstract, material and metaphorical. This initial insight led him to his thematic trialect of space. He argued social space constitutes a spatial practice (i.e. perceived space), representations of space (i.e. conceived space), and spaces of representation (i.e. lived space). Lefebvre's attention to the conditions of social space has influenced the theoretical writings of Jean Baudrillard, Michel de Certeau, David Harvey, and Edward Soja, among others. For urban designers, Lefebvre's ideas have been popular, often by those opposed to large-scale designs that they perceive as inflexible or oppressive.

His theories continue to be relevant in the neoliberal era among planning and design scholars.

Geographer Edward Soja describes Lefebvre's thematic trialect while arriving at his concept of thirdspace in Los Angeles.[11] Soja states that Lefebvre's perceived space refers to the materialized, socially produced, empirical space of everyday life. Perceived space is a process of producing the material form of social spatiality; therefore, it is both a medium and an outcome of human activity, behavior, and experience. Conceived space, on the other hand, references the abstraction of perceived and lived space by scientists, planners, and urbanists who impose an order or vision. Conceived space is constituted via the control over the production of spatial knowledge. Finally, lived space encompasses both perceived and conceived space while drawing attention to the inhabitants and users that partake in a social struggle. As Soja describes, "lived spaces of representation are the terrain for the generation of counterspaces, space so resistant to the dominant order arising precisely from their subordinate, peripheral or marginalized positioning" (68). Perceived space considers the materials (i.e. what is seen) and conceived space considers the abstract (i.e. what is thought), while lived space is embodied (i.e. what is felt).

From 1957 to 1972, the Situationist International was a small group of revolutionaries that shaped the May 1968 French uprisings and published multiple critiques of advanced capitalism such as the *Report on the Construction of Situations*[12] and *The Society of the Spectacle*[13] by Guy Debord, and *The Revolution of Everyday Life*[14] by Raoul Vaneigem, among others. These texts rearticulated classical Marxist theories to address the decline of human life for individuals and society. With roots in Dada and Surrealism, the situationists asserted that the rise of a consumer society and technological progress caused social and psychological alienation and profoundly altered work, leisure, and play. Their concept of the derive (i.e. drifting) through the urban environment, became a useful tool to explain a human-centered psychogeographical experience whereby users navigate without intention and let the forgotten, discarded, or marginalized aspects of the urban environment lead them on a journey. The derive disregards purposeful agendas and allows humans to reclaim their experience with the agency as actors in the city.

Everyday Urbanism,[15] an approach introduced by Margaret Crawford, John Chase, and John Kaliski, specifically eschews interest in aesthetics in favor of an understanding of everyday life. This empirical approach epitomizes a bottom-up approach, looking at the contingent and episodic as the important measure of space in the city, and explicitly acknowledges the work of Lefebvre and Debord. This approach is also notable for its interest in specific times and places and its embrace of the temporary and the short-lived; street markets, food trucks, and street art are their preferred sites. Most of the investigations have centered on ethnic minorities, who appropriate marginal space for new occupations.[16] Everyday urbanism has as its goal "a call to action" by "working along with, on top of or after" rather than a complete transformation.

In *Skateboarding, Space and the City: Architecture and the Body*[17] Iain Borden applies Lefebvre's analytic framework toward understanding the emergence of skateboarding, seeking to understand how "...space is part of a dialectical process between itself and human agency". Using an archive of Skater magazines as primary sources, he traces the evolution from an activity for bored teenagers to a Southern California phenomenon to a worldwide urban subculture. He describes both a set of inventions (roller skates bolted to boards) and appropriations (empty swimming pools as the first skating sites). Borden is particularly interested in the interplay between space and the body of the skater, as their tricks literally diagram their occupation of space. The development of "tricks" in the sport and the efforts of public officials to block them are a ballet of top-down and bottom-up agency.

The general thrust of these theories highlights the dynamics of the socially constructed landscape of cities and offers a critique of ideologies that exclude or neglect social diversity in the city. In doing so, they offer a perspective of cities where performance and citizen participation is the DNA of the city.

Toward a Theory of the Contemporary City

Faith in the arrival of digital technology and belief in the socially constructed reality offer no simple resolution leading to a comprehensive theory. However, theories in other fields including geography, sociology, and science and technology studies offer approaches that address the dichotomy of the social and the technological and can form the basis for understanding social media in the contemporary city.

Bruno Latour is a leading voice in science and technology studies who aims to understand the social setting and the implications of science. He was educated in philosophy and sociology and taught at the Ecoles des Mines and Sciences Po in Paris.

Latour's early work was notable for ethnographic studies of the behavior of scientists in laboratory settings, most notably published in the book *Laboratory Life*[18] after a year-long residency at the Salk Institute. He and his co-authors' controversial claims that science is socially constructed within the laboratory and that scientists often must decide which finding to accept and which to ignore establish the idea that science and the social setting are tightly intertwined.

Latour later collaborated with Michel Callon and sociologist John Law to develop Actor-Network Theory (ANT), an agent-based approach to social theory and research in the fields of sociology and technology.[19] ANT can be described as a "material semiotic" method, which means that it analyzes relationships that are material (between things) and semiotic (between concepts) and argues that many relations may be both material and semiotic. For ANT, people and objects are treated as part of a network in an attempt to understand the process of technological innovation and adoption. In the 1990s, ANT became a popular analysis tool for a wide range of fields. Today it is a widely used approach to study how people and objects interact with each other at multiple scales such as small technological innovations (i.e. a door-closer), and in larger, more impacting innovations (i.e. mobile phones).

More specifically, ANT seeks to overcome the subject-object divide and the distinction between the social and the natural worlds for a network to form and a technology to be enacted. Translations in ANT describe continual displacements and transformations of subjects and objects and the insecurity and fragility of change and their susceptibility to failure. Technology needs to be easy to use, accessible, and take cultural and ethnic groups into account. The technological and social aspects of a community dictate how the technology will or will not integrate into society. People tend to rely on others to influence the success or failure of a product. The communities' views on the technology are based on their background, sharing the technology, or refusing it. Once a technology becomes stable in a community, it is translated in the actor-network sense of becoming a black box that becomes reliable and unremarkable.

In a playful article written under a pseudonym,[20] Latour imagines a series of human and technical substitutions that led to the door closer at the front door of the sociology department of an imaginary university. First, the movement through the wall consists of a hole in the wall, which leads to the development of a door to separate the outdoor and the indoor. To solve the problem of keeping the door closed, a doorman was introduced. However, no matter how well he does his job, there could be a day where the doorman is ill or on a break, which brings forth the invention of the door closer.

Translation, an important aspect of ANT, refers to the change in the interrelationship of actants in a network. ANT assumes that the shifting roles of these actants should be the main focus of study rather than any preassigned categories. This is a constructivist rather than an essentialist method consistent with a bottom-up rather than a top-down approach. One example of such a translation is the shift from movie theaters to Blockbusters stores to streaming video. The physical objects, technologies, humans involved, and their interrelationships have reconfigured over the last 60 years.

Studies of the process of translation are understood to involve four stages: problematization, interessement, enrollment, and mobilization.[21] While this could be a very lengthy and difficult process, a defining aspect of social media is the very rapid deployment and adoption. For example, in the first three years after its introduction, Instagram had 200 million users and in seven-and-half years had one billion. Translations based on social media have become widespread and nearly instantaneous.

Stephen Graham is a British geographer and urbanist who has published extensively on the contemporary transformations of the city. In Graham's earliest work he explicitly addresses the conceptual relationship of space and information technology in *The end of geography or the explosion of place? Conceptualizing space, place and information technology.*[22] In 2001, he became well-known for his book *Splintering Urbanism*[23] which examines the emergence of private infrastructure systems and the resulting uneven distribution of access to information and communication technology. *Telecommunications and the city*[24] provides an overview of the relationship between communication systems and city development and management. More recently, *Cities Under Siege: The New Military Urbanism*[25] discusses the effect of military and security forces on cities across the globe.

Graham's work debunks technological determinism that claims communication technologies substitute or transcend physical movement or the city itself. This substitutional argument assumes that technological advances will replace existing spatial structures and materiality, substituting physical propinquity with electronic delivery. The city will be rendered unnecessary and replaced by instant communication or the creation of a "mirror world". Graham suggests that the consequence of these narrow ideas about the end of geography tends to be that we don't pay enough attention to the hidden substrata of technologized, materialized infrastructure. He suggests that we need to take a closer look at the materiality of communication systems that are powered by vast amounts of electricity and are materialized through unimaginably complex systems of fibers and servers and satellite dishes, which are sedimented into the landscape and the city as a means to overcome spaces and times.[26] Ultimately, globalized urbanization continues to gather force rather than being supplanted.

In *Splintering Urbanism*, Graham proposes a co-evolution of geographic and electronic space in what he calls socio-technical assemblages drawing from the work of ANT theorists and the post-structural philosophies of Gilles Deleuze. He argues that complex articulations between place-based and telemediated relationships have emerged that are simultaneously social and technical. The richness and particularity of human communications in a specific place are poorly captured by electronic media. Rather, localized, physical relationships and technological relationships are complimentary just as urban space and telecommunications continue to shape each other recursively. Networks vary in their flexibility and ability to adapt to new spatial arrangements, for example, delivery and transportation networks. The creation of favored locations within the city creates privileged neighborhoods and information "deserts".

Graham provocatively suggests that it is impossible to live in a modern, urban life that is not profoundly based on the whole multiplicity of assemblages that blur the social and the technical on cyborg-like assemblages.[27] In this view, space is neither dissolved nor transcended but constructed through interactions and replacements with technical and human "actants" in virtual and physical contexts. In order to capture the way assemblages are meshed into the politics of space and the city, you have to look at it as a dynamic process of urban life continually coming into being, in a profound sense of ongoing process and dynamism[28].

Manuel Castells is a Spanish sociologist and city planner who is known for his work on the influence of information technology on urban planning, production of space, and power. His work *The Informational City: Economic Restructuring and Urban Development*,[29] followed by *The Rise of the Network Society*[30] summarizes changes to the regional and global economies caused by the widespread adoption of information technology in the 1970s and 1980s. In this period of economic restructuring, a firm's competitive advantage is based on their knowledge and access to information technology that concentrates in cities and their ability to build international networks. This period of globalized capital propelled by the decentralizing force of networks has led to the space of flows in which the affordances of distant synchronous interaction supersedes the meaning of places and fragments the workforce. Although at one level it argues for the triumph of technological networks over propinquity, Castells upholds that the physical location of financial firms in urban centers is an important part of the contemporary setting.

Building upon his Marxist urban sociological roots, Castells also analyzes social movements over time beginning with *The City and the Grassroots*,[31] *Communication Power*,[32] and more recently, in *Networks of Outrage and Hope*.[33] These books explicitly address Castells's concerns with political protests arising in opposition to dominant political power which he calls networked social movements. He examines trade unionism, squatter movements, and urban riots in a number of countries including Tunisia, Egypt, Spain, France, and the United States. He views uprisings as a collective practice able to produce changes in urban systems, local culture, and political institutions in contradiction to dominant social interests. He argues communication networks are central sources of power in political movements where users have agency and decision-making is framed by emotional responses to information. Communication is not only about the means to transmit messages but more profoundly technology plays a critical role in the experience of social movements and situates virtual and physical realms as mutually dependent.

Castells identifies three critical operative components of these movements that are important to this research. First is the creation of community, an idea of togetherness as a means to overcome fear. This is often forged at the local level initiated by direct contact and trust on a personal level. Second, the urban spaces of these movements are often charged with symbolic power. Sometimes they symbolize the power of state or financial institutions that are the focus of these social movements. Often, they are the sites of important previous political action, whether it be the Easter Rising in Dublin or the Paris Commune of 1848. Third, a social network brings together the beliefs and values of a local community. Social media networks are capable of connecting large numbers of people almost instantaneously toward a common goal. It is a method to directly create continuous interaction, discussion, action. Castells is clear that all three are necessary and mutually supportive; community, urban space, and social media.

Case Studies

The ten chapters at the center of this book are a series of case studies focused on the intersection of the local community, urban space, and social media in the contemporary city. The choice of case studies varies in scale, duration, intended audience, location, and use of urban space. Some are in-depth studies of all three factors, while others focus more narrowly on one aspect. If these studies are interrogations, some are intensive and wide-ranging while others ask a narrower set of questions.

Political protests are the focus of the first set of case studies. They offer a range of political and cultural settings, issues with security forces, and duration.

Chapter 3 is an analysis of a Black Lives Matter protest following the shooting of Keith Lamont Scott on September 10–12 in Charlotte, NC. The change in strategy and location over this three-day period was partly spontaneous and partly organized by the local community. The emergence of two distinct online communities is also notable.

Chapter 4 deals with Hong Kong concerning the public protests that lasted for the entire year of 2020 and continues in a smaller form as this book is being written. These protests are notable for their shifting location in urban space, their use of a wide variety of social media platforms to avoid police surveillance, and the strength and resilience of the local community that organized and guided the protests to new and reoccuring locations.

Chapter 5 focuses on the Women's March in Washington DC on January 21, 2017. This event is notable both for its location at sites of national importance, as well as for the importance of social media in creating and sustaining a shared sense of community through posting and reposting photos. Communications using social media began during the planning phase using Facebook, later transitioning to multiple forms of social media including prominently Instagram.

The next group studies the use of social media for informal commerce in private and marginal urban spaces that were adapted for their temporary use.

Chapter 6 is a study of mobile food trucks in Charlotte, North Carolina in 2015, over three months. The ability of food trucks to choose and change locations and the methods that they use to schedule and announce their intentions are salient features of this chapter. A detailed analysis of food truck locations and interviews with owners provide a detailed picture of the interplay between the local community, urban space, and social media.

Chapter 7 is a study of sidewalk vendors in Iran who dynamically navigate between urban space and social media to sell goods over three months in the summer of 2019. A thematic analysis of vendor interviews and an exploratory analysis of their social media accounts shows how they make their living in both social media and urban space.

Chapter 8 focuses on patrons' immersive experiences of underground restaurants. The use of social media is not often extensive or vital to notify patrons of an upcoming dinner. Owners instead publicize through word-of-mouth, websites, or photographs. This chapter analyzes the images and posts on Instagram to understand how chefs and patrons construct photographic experiences and expectations of dining events.

Art and culture are the focus of the next group. These events occur during a defined period of time and use urban space as a stage and social media as an organizing strategy to publicize the festival as it unfolds.

Chapter 9 is focused on Banksy installations in Bristol, England and a month-long residency in New York City in 2020. The clear beginning to these events is important, as well as the connection to the local community and a distant, digital audience. Secondary events related to the relocation, preservation, or erasure of the installations play out over a longer time frame. The engagement of social media is prominent in all the installations although it shifts its focus from discovery to announcement.

Chapter 10 centers on the Burning Man festival, a yearly ten-day event gathering held in the Nevada desert at Black Rock City, a temporary urban encampment of 60,000. Because of the cancellation of the festival due to the Covid19 pandemic in 2020, it is possible to study how the prominent space of all Burning Man festivals is reconfigured in virtual form.

The final group of case studies is focused on mass shootings and the involvement of social media before, during, and after these events.

Chapter 11 focuses on the massacre at the Al Noor Mosque in Christchurch, New Zealand on March 15, 2019. This event has a dual reading, one based on the white supremacist online community connected to the assassin and the other on the reaction of those sympathetic to the victims and values of diversity and inclusion. The discussion on social media reinforces the role of the mosque as a location for solidarity and healing that was of central importance to the local community.

Chapter 12 focuses on the shootings at the Pulse Nightclub in Orlando, Florida in 2016. The outpouring on social media as a consequence of this event was organized and understood topically. Memorials in many forms are an important part of the process of grieving and remembering. Memorials took the form of online tributes, ad hoc memorials at the nightclub, a temporary memorial fence, and plans for a new museum structure on the site. Through topic modelling and image analysis the chapter studies the effect and reaction to these memorials on social media.

Each chapter begins with an explanation of the historical and spatial settings, including the precise timing of the event and their important urban locations. An ethnographic understanding of the meaning intended by the organizers and participants is also vital to each case study. Beginning with a survey of scholarly research about each event provides an understanding from a wide range of disciplines. Additional insight can be accomplished using accepted methods from urban design and planning and anthropology, but analyzing mobile social media data requires new methods. Therefore, in each case, there is an explanation of the methodological choices. This is based on a conviction that there is no single method that works for all situations, and that fluency with data analytic methods will be required to fit each particular case.

A sample configuration diagram of these relationships is presented in Illustration 1.2. Nodes signify local community, urban space, or social media. Edges of the graph indicate actions or meanings and can be uni- or bidirectional. They can also show the direction of influence and the variation over time on each edge. The work by Arsenault and Castells on the

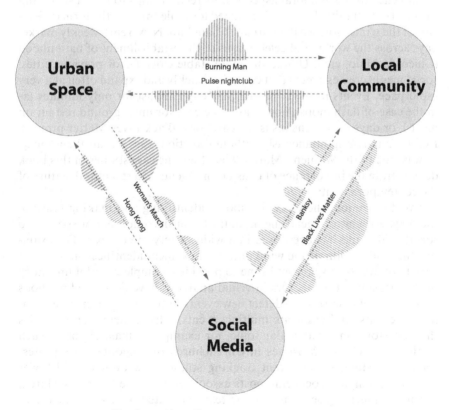

ILLUSTRATION 1.2 Configuration diagram.

complex constellation of actors, organizations, and influences in the misinformation campaign leading to the Iraq invasion was an inspiration for this visualization.[34]

The local community is the shared perspectives of people, friends, and families who meet each other in physical space or private media such as phone calls or text messaging. The goal is to study how people connect with each other and distribute ideas and concerns. This will allow an understanding of the motives and how they navigate urban space and social media.

Urban space includes the locations where both the local community and the social networks select and focus their activity. Both symbolic and functional urban spaces are important locations where communities form, demands are made, and focused activities occur.

Social media includes sites on mobile social media where information and thoughts are shared, rapid communication is made, and shared activity is planned. While some communication on social media is between people who know each other, much of it is anonymous, public, and widely distributed.

Because social media provides a way to communicate in real-time, it gives prominence to the temporal aspects of its relationships to urban space and the local community. Events have existed for millennials in the form of festivals of the winter solstice (Christmas, Chanukah, New Year), weekly market days across the world, and celebrations of the establishment of nationhood (Cinco de Maio, etc.). Unlike the more stable calendar of previous cities, social media creates events that can arise quickly and expand often at a very rapid pace. Events in the contemporary city can last for only minutes (as in the case of flash mobs), hours (as is the case for underground restaurant meals), or days or months (as is the case for a Black Lives Matter protest). Events can arise spontaneously without direction from a central authority, as was true of the Women's March. Therefore, the case studies in this book demonstrate the importance of considering events as a structural feature of the contemporary city.

Event detection, therefore, is vital to identifying and making sense of the huge amount of social media data. Event detection is not specific to social media and has been used in a wide variety of contexts. For example, the study of hurricane wind and eye movement identifies and tracks a variety of events in space and time to provide a complete, unfolding story of the disaster.[35] Researchers in visual analytics have developed methods useful in diverse settings to detect new events, track events over time, summarize events, and associate multiple events.[36] Event detection identifies the first story on a particular topic, for example, a mass shooting such as the Christchurch Mosques in Christchurch or a spontaneous protest such as in Hong Kong. Event tracking studies how events unfold over time, for example, reoccuring posts associated with the Women's March. Event summarizing seeks to use the text associated with social media to identify the content of the event. Event association seeks to understand the relationship of an event to other factors, which in this book focused on connection to urban space and the local community. An example is the identification of local and nonlocal online communities during the Black Lives Matter protest.

Thus, the forms and location of urban space, the nature and composition of the local community, and the influence of social media shift over time. These case studies reveal differing roles of social media, local communities, and urban space, as well as a critical role for the timing, locations, and reactions to events. The roles of live protesters, media representation, and urban space can be linked in multiple ways across different domains. Borrowing from ANT theory, it is characterized by the translation of responsibilities between "actants",[37] some live, some digital media, and some spatial shifting (often rapidly) over time.

In the 15th century, when new media arrived in the form of movable type and widespread literacy, it made promises similar to those now offered for digital media. Victor Hugo assessed the shift in power from the clerical elite to a wide and diffuse reading audience.[38] "The Book will destroy the Edifice" he proclaims, surely a move from authoritarianism to democracy. But in the end, Hugo calls it "the second tower of Babel", a society no less prone to manipulation and questionable actions than that which preceded it.

The final chapter of this book begins with a summary of the findings of the case studies with a focus on the translation of existing patterns of urban space, local community, and social media. The goal is to create an understanding of how the introduction of social media has shifted rather than replaced the roles of the local community and urban space. Then the advantages and challenges of using social media as an analytic method are discussed, including the need for a general method of exploring and understanding massive amounts of social media data. This discussion includes the need for methods that are more generally applicable than the bespoke investigation that is typical of current analysis; visual analytic systems are identified as a promising alternative. Finally, the ethical and political implications of the ready accessibility of data as a regular aspect of contemporary cities are discussed. It would be foolish not to notice the advances and affordances of this technology, but it would be equally jejune to think it is without costs. Issues of privacy and surveillance, conspiracy and resistance, racial animus and shared humanity, collective intelligence, and monetization are not solvable by technical means. The shape and nature of these challenges and audiences have a particular form and impact on the contemporary city.

Notes

1 Krikorian, Raffi. (2013). "New tweets per second record, and how". *Twitter Engineering Blog, 16*. blog.twitter.com.

2 Moravic, Hans. (1998). "The senses have no future". In J. Beckmann (Ed.), *The Virtual Dimension: Architecture, Representation, and Crash Culture* (pp. 85–95). Princeton, NJ: Princeton Architectural Press.

3 Mitchell, William. (n.d.). *Etopia: 'Urban Life, Jim--but Not as We Know it' (1999).* Cambridge: MIT Press.
   Mitchell, William J. (1996). *City of Bits: Space, Place, and the Infobahn.* Cambridge: MIT Press.
   Mitchell, William J. (2004). *Me++: The Cyborg Self and the Networked City.* Cambridge: MIT Press.

4 Page 3, Mitchell, William. (n.d.). *Etopia: 'Urban Life, Jim--but Not as We Know It'. (1999).* Cambridge: MIT Press.

5 Nam, Taewoo, & Theresa A. Pardo. (2011). "Conceptualizing smart city with dimensions of technology, people, and institutions". *Proceedings of the 12th Annual International Digital Government Research Conference: Digital Government Innovation in Challenging Times.*

6 D'Andrea, Eleonora, Pietro Ducange, Beatrice Lazzerini, & Francesco Marcelloni. (2015). "Real-time detection of traffic from Twitter stream analysis". *IEEE Transactions on Intelligent Transportation Systems.* https://doi.org/10.1109/tits.2015.2404431.

7 Ratti, Carlo & Matthew Claudel. (2016). *The City of Tomorrow: Sensors, Networks, Hackers, and the Future of Urban Life.* New Haven, CT: Yale University Press.

8 Lefebvre, Henri. (1991a). *Critique of Everyday Life - Volume 1: Introduction,* trans. John Moore, London and New York: Verso, [first published *Critique de la vie quotidienne I: Introduction,* Paris: Grasset, 1947].

9 Lefebvre, Henri. (1991b). *The Production of Space,* trans. D Nicholson-Smith, Oxford, UK: Blackwell [first published *La production de l'espace,* 1974].

10 Lefebvre, Henri. (1996). *Writings on Cities* (Eleonore Kofman & Elizabeth Lebas, Eds. & Trans.). Oxford: Blackwell.

11 Soja, Edward W. (1996). *Thirdspace: Journeys to Los Angeles and Other Real-and-imagined Places.* Cambridge: Blackwell Publishers.

12 Debord, Guy. (2006). "Report on the construction of situations". In *Situationist International Anthology.* Berkeley, CA: Bureau of Public Secrets, translated by Ken Knabb. [first published as "Rapport sur la construction des situations", 1957].

13 Debord, Guy. (1994). *The Society of the Spectacle.* New York: Zone Books. [first published as *La société du spectacle by Buchet-Chastel,* in French, 1967].

14 Vaneigem, Raoul. (1972). *The Revolution of Everyday Life.* Practical paradise Publications. [first published as *Traité de savoir-vivre à l'usage des jeunes générations* by Gallimard, in French, 1967].

15 Chase, John, Margaret Crawford, & John Kaliski (eds.). (1999). *Everyday Urbanism.* New York: Monacelli Press.

16 Scholarship addressing everyday urbanism includes: Mukhija, Vinit and Anastasia Loukaitou-Sideris (eds.). (2014). *The Informal American City: Beyond Taco Trucks and Day Labor.* Cambridge, MA: The MIT Press; Hou, Jeffery (Ed). (2010). *Insurgent Public Space: Guerilla Urbanism and the Remaking of Contemporary Cities.* New York: Routledge; Lydon, Mike & Anthony Garcia (eds.). (2014). *Tactical Urbanism.* Washington, DC: Island Press.

17 Borden, Iain. (2001). *Skateboarding, Space and the City: Architecture and the Body.* Oxford: Berg.

18 Latour, Bruno & Steve Woolgar. (1986) [1979]. *Laboratory Life: The Construction of Scientific Facts.* Princeton, NJ: Princeton University Press.

19 Latour, Bruno. (1996). "On actor-network theory: A few clarifications." *Soziale Welt, 47*(4), 369–381.

20 Latour, Bruno [Johnson, Jim]. (1988). "Mixing humans and nonhumans together: The sociology of a door-closer". *Social Problems, 35*(3), 298–310.
21 Callon, Michel. (1986). "Some elements of a sociology of translation: Domestication of the scallops and the fishermen of St Brieuc Bay." In John Law (Ed.), *Power, Action and Belief: A New Sociology of Knowledge* (pp. 196–233). London: Routledge & Kegan Paul.
22 Graham, Stephen. (1998). "The end of geography or the explosion of place? Conceptualizing space, place and information technology." *Progress in Human Geography, 22*(2), 165–185.
23 Graham, Stephen & Simon Marvin. (2001). *Splintering Urbanism: Networked Infrastructures, Technological Mobilities and the Urban Condition*. Oxfordshire: Psychology Press.
24 Graham, Steve & Simon Marvin. (2002). *Telecommunications and the City: Electronic Spaces, Urban Places*. Oxfordshire: Routledge.
25 Graham, Stephen. (2011). *Cities under Siege: The New Military Urbanism*. Brooklyn, NY: Verso Books.
26 Farías, Ignacio. (2010). "Interview with Stephen Graham". In Ignacio Farías & Thomas Bender (Eds.), *Urban Assemblages: How Actor-Network Theory Changes Urban Studies* (pp. 197–205). London: Routledge.
27 Cyborg-like assemblages continually connect human life with distant times and spaces through the metabolism of the body, through the metabolisms of excretion, consumption, food, energy, production, transportation, and so on (2010:198). Based on a relational understanding of the technological and social worlds, this leads to multiple cyberspaces reacting to heterogeneous forces.
28 Graham, Stephen. (2014). "Interview with Stephen Graham". In I. Farias and T. Bender (Eds.), *Urban Assemblages: How. Actor-Network Theory Changes Urban Studies* (pp. 197–205). London: Routledge.
29 Castells, Manuel. (1989). *The Informational City: Information Technology, Economic Restructuring and the Urban-Regional Process*. Oxford: Blackwell.
30 Castells, Manuel. (1996, second edition, 2009). *The Rise of the Network Society, the Information Age: Economy, Society and Culture Vol. I*. Malden, MA; Oxford: Blackwell.
31 Castells, Manuel. (1983). *The City and the Grassroots: A Cross-cultural Theory of Urban Social Movements*. Berkeley: University of California Press.
32 Castells, Manuel. (2009). *Communication Power*. Oxford: Oxford University Press.
33 Castells, Manuel. (2015). *Networks of Outrage and Hope: Social Movements in the Internet Age*. John Wiley & Sons.
34 Arsenault, Amelia & Manuel Castells. (2006). "Conquering the minds, Conquering Iraq: The social production of misinformation in the United States – a case study." *Information, Communication & Society*. https://doi.org/10.1080/13691180600751256
35 https://viscenter.uncc.edu/research/diem-techtransfer/event-structure-analysis-hurricane-wind-and-eye-movement.
36 Dou, Wenwen, Xiaoyu Wang, William Ribarsky, & Michelle Zhou. (October 2012). "Event detection in social media data." In *IEEE VisWeek Workshop on Interactive Visual Text Analytics-Task Driven Analytics of Social Media Content* (pp. 971–980). Seattle, WA: IEEE.
   Dou, Wenwen, Xiaoyu Wang, Drew Skau, William Ribarsky, & Michelle X. Zhou. (October 2012). "Leadline: Interactive visual analysis of text data through event identification and exploration". In *2012 IEEE Conference on Visual Analytics Science and Technology (VAST)* (pp. 93–102). IEEE.

37 See for example Bruno Latour's pseudonymous description of the translation of human and machine actants in Johnson, Jim. (1988). "Mixing humans and nonhumans together: The sociology of a door-closer." *Social Problems, 35*(3), 298–310.

38 Wren, K. (1993). *Introduction Victor Hugo: The Hunchback of Notre-Dame.* London: Wordsworth Classics.

2

METHODOLOGY

In the introduction a case was made for the importance of considering urban space, the local community, and social media as the three critical aspects of a network that describes the contemporary city. This chapter will discuss important methods associated with each that form the foundation for the case studies that follow.

Because urban space is the oldest and most settled method, relatively more focus is on emerging qualitative and quantitative methods for investigating local communities and social media.

Urban Space

Site Analysis and Documentation

Traditionally, site analysis in the fields of urban design, architecture, and planning involves systematically looking at the physical and social surroundings of a location to find out how an environment got to be the way it is, what decisions its designers and builders made about the place, how people actually use it, how they feel about their surroundings, and generally, how that particular environment meets the needs of its users.[1] Using annotated diagrams, drawings, sketches, photographs, and counting techniques a researcher can document a setting unobtrusively by observation. Site analysis typically begins with a diagrammatic plan of the setting that is used to note where objects and buildings are located, physical traces of human activity throughout the site, and points of access to and from the site or building structure. Photographs of the site features and social activity are useful to capture details of the setting at a particular moment in time that can be studied after the visit. Counting may involve tallying the presence of objects or types of human activity (i.e. sitting, walking, talking). In addition to capturing the physical organization

DOI: 10.4324/9781003026068-2

of objects in the space and traces of human activity, a site researcher will also survey features of the natural landscape (i.e. temperature, precipitation, sun orientation, ground slopes, winds, presence of water, seasonal changes), adaptations of the site that are uncharacteristic of their intended use, official and unofficial messages found in signage and written markings, and hints of personalization from objects that display characteristics of oneself. Using this information, a researcher can describe the qualities of the setting and how people actually use it. The information will reveal shortcomings in the current setting that can be used as evidence to make improvements.

Analysis Using Social Media

Social media is a more recent tool used to analyze the characteristics of urban space. Researchers using mapping techniques and computational analysis can study social media information to understand human behavior and their sentiments about various urban spaces. For instance, the geolocation information in social media messages provides a way to accurately locate social media users in a landscape. Mapping locations over time can reveal insights about patterns of use or abandonment in various sectors of a city. Traffic planners and everyday citizens use mapping applications that source the location of social media users to alleviate traffic congestion.

The content of social media messages can provide information about the experiential qualities of a place and the users' comfort and preferences in the environment. Studies addressing user sentiment of green space,[2] bus systems,[3] and commerce reveal negative emotional reactions that can lead to problem identification and better design and planning solutions.

Social media messages can also provide information on unexpected, real-time events taking place in cities to allocate emergency resources, such as car accidents, flooding, or wildfires.

For purposes of this research, we explore a variety of social media content depending on the nature of the event. Data frequency patterns, users' location, or personal sentiments are all valuable points of investigation.

Local Community

This section focuses on qualitative methods and data from a background in ethnography, grounded theory, and thematic analysis as a framework for guiding the analysis of formal interviews, surveys, print articles, blogs, and scholarship in conditions of uncertainty.

Grounded Theory

In the mid-century, Barney G. Glaser and Anselm L. Strauss's sociological *grounded theory* challenged the mainstream quantitative research that seldom led to new theory construction.[4] By challenging these positivist conceptions that stressed objectivity, generality, replication of research, and falsification, grounded theory analysis allows the researcher to participate in the creation of theory and merge facts with values. *Grounded theory* allows for the inductive development of theories that speak to the reality of on-the-ground occurrences, as opposed to deducing testable hypotheses from existing theories. Grounded theory merges two contrasting traditions. Glaser's background in quantitative training is reflected in the Columbia University positivism. Glaser's interests in empiricism, rigorous codified methods, emphasis on emergent discoveries echo approaches to quantitative analysis. Whereas, Strauss's Chicago School heritage in grounded theory brings notions of human agency, emergent processes, social and subjective meanings, problem-solving practices, and open-ended strategies.[5]

Over time grounded theory evolved into two general paradigms: constructivists and objectivists. Constructivists place priority on the phenomena of study and see both data and analysis as created from the shared experiences of researcher and participants, whereas objectivists, the researcher remains separate and distant from research participants and realities.[6] The dual roots and multiple approaches to grounded theory allow the practice to remain popular for decades with a number of interpretations and applications.

Kathy Charmaz, a well-known scholar in qualitative methods, clearly summarizes the key components of grounded theory practice:

1 Simultaneous involvement in data collection and analysis
2 Constructing analytic codes and categories from data, not from preconceived logically deduced hypotheses
3 Using the constant comparative method, which involves making comparisons during each stage of the analysis
4 Advancing theory development during each step of data collection and analysis
5 Memo-writing to elaborate categories, specify their properties, define relationships between categories, and identify gaps
6 Sampling aimed toward theory construction, not for population representativeness
7 Conducting the literature review after developing an independent analysis.[7]

Grounded theory provides a broad framework to address a qualitative analysis of interviews, field research, and news reports. Important aspects of the constructivists' assumptions include the observer as an active participant in the data creation and analysis and its bottom-up approach to theory development.

Thematic Analysis

In qualitative psychology, thematic analysis is a popular adaptation of grounded theory. Despite its lack of acknowledgment, thematic analysis is widely used for its practical use of coding. Thematic analysis is a method for identifying, analyzing, and reporting patterns (themes) within data.[8] It requires searching across a dataset (i.e. interviews, texts) to find repeated patterns of meaning.

Thematic analysis has been viewed as a process performed "within" an analytic tradition such as grounded theory or interpretative phenomenological analysis, however it has a few unique characteristics. The method is not wedded to any preexisting theoretical framework, which allows it to be applied to different topics easily. Second, thematic analysis does not attempt to quantify patterns in the data, rather a theme is determined to be significant by its relation to the research question under study. Last, thematic analysis does not require specific theoretical or technical knowledge to perform, which allows it to be a relatively easy and quick method to learn.

Themes, developed by the researcher, capture something important about the data and represent some level of patterned response or meaning within the dataset. Themes can be identified in one of two primary ways: inductive or theoretical. Inductive analysis is a process of coding the data without fitting it into a preexisting coding frame or the researcher's analytic preconceptions. In other words, it is data-driven. Similar to grounded theory, themes are identified as strongly linked to the data themselves. In contrast, a theoretical thematic analysis is driven by the researcher's theoretical or analytic interest and is more explicitly analyst-driven. Therefore, you can either code for a specific research question (which maps onto the theoretical approach) or the specific research question can evolve through the coding process (inductive approach) (84). In our work, we approach identifying themes using a theoretical approach to find information related to the ways social media performs in the social, cultural, and economic activities taking place in the city.

The steps of thematic analysis include:

1 Familiarizing yourself with your data: Transcribing data, reading and re-reading data, noting down initial ideas.
2 Generating initial codes: Coding interesting features of the data in a systematic fashion across the entire dataset.
3 Searching for themes: Collating codes into potential themes.
4 Reviewing themes: Checking if the themes work with the coded extracts and the entire dataset, generating a thematic map of the analysis.
5 Defining and naming themes: Ongoing analysis to refine the specifics of each theme, and the overall story of the analysis, generating clear definitions and names for each theme.
6 Producing the report: Selection of vivid, compelling extract examples, final analysis of selected extracts, relating the analysis to the research question and literature, producing a scholarly report.

Thematic analysis is useful for its wide applicability across disciplines and clear steps of implementation.

Semi-Structured Interviews

Interviews provide a wealth of information about participants' daily work habits, decision-making, preferences, and purposes. Semi-structured interviews allow probing, or prompting the participant based on their responses. Interviews consist of open-ended questions derived from curiosities about the basic processes of participants' actions during an event or their involvement with social media. In-person and telephone interviews were conducted on food vendors in the United States and Iran, graffiti artists, and supper club chefs.

Participant Surveys

Participant surveys provide a quantitative or numeric description of trends, attitudes, or opinions of a population by studying a sample of that population.[9] The purpose of our surveys conducted on food vendors is to understand better the trends among their daily practices when conducting their business and how social media alters or enhances the ways they conduct or receive business. Surveys were captured at one point in time, using anonymous online forms.

Virtual Ethnography

Virtual ethnography or online ethnography, familiar to the field of information science, allows researchers to both observe and participate in online communities.[10] The significance of virtual ethnography emerges from the active use of social media in the daily practices of research participants. In this research, the exchange of real-time information that occurs online becomes equally important as the physical social settings where their practices take place. For example, in our chapter that discusses mobile food vending, the vendors are immersed as participants in social media platforms such as Twitter, Facebook, and Instagram. By observing and participating in these platforms we are able to understand how online communication affects the organization and performance of social relationships in time and space. Opposed to understanding how a virtual experience might be different from "the real", virtual ethnography allows researchers to become active in the participants' online social worlds.

Using a mixed methods approach provides a way to sufficiently explain the contemporary and rapidly evolving nature of our case studies. In addition to grounded theory and virtual ethnography methodologies, we explored a variety of quantitative techniques to analyze online social media data.

Social Media

For each social media message, metadata typically identify the time, sender, and sometimes the topic (e.g. hashtags), but the content of the message can vary widely and has no inherent organization. Social media generates billions of messages, well beyond the ability of any individual or group of humans to read and organize. Understanding and making sense of the enormous archives of messages requires specialized data analytic techniques such as topic modeling, named entity extraction, and structural analysis. The quantitative modeling and data collection section of this chapter explains these issues and each chapter includes a discussion of the modeling techniques used for each example.

Social media has many overlapping platforms, which vary by region. There is not a single repository of data; Twitter may be a rich source of data in New York, but in Hong Kong or Tehran, Telegram may be much more popular. Differences are partly due to official censorship, partly to cultural preference, and partly to differing levels of privacy. Throughout this book, Twitter, Instagram, Telegram, Reddit, Facebook, and other social media are all used as sources of data. Each chapter includes a discussion of the social media channels used by participants.

The spontaneous and often insurgent events in this book are particularly suited to the affordances of social media. Social media allows instant multiuser communication and the rapid emergence of online communities of interest and is at best lightly curated, which can be important for politically or socially marginalized groups. This will emerge as an important and not unambivalent aspect of the use of social media.

While it is possible for social media to be directed top down (e.g. political party campaigning), it is also well suited to the growth of bottom-up informal activities. This book focuses on the spontaneous emergence of such online communities, either as protests, quasi-legal commerce, or terrorist attacks. Even those examples with top-down planning strategies such as Banksy or the Burning Man are careful to provide a participatory role for attendees and a prominent role for the participants.

One issue regarding social media data about unplanned events started by bottom-up communities is that they usually have only loosely articulated and emergent intentions. Data cannot provide a complete understanding of all participants' diverse intentions or reactions to an event. This means that it is not possible to assume the structure of the data. For this reason, the qualitative methods may be better at defining the motivations for the event, how participants negotiate challenges or take advantage of opportunities, and deeper clues into their emotional states of mind. Although qualitative information is rich in depth, the quantity of the information is more limited than social media data and better analyzed using thematic analysis.

Social media usage within cities has a fascinating and unique feature. As people live and move through cities and social media, they leave huge amounts of rich but complex data that include information about topics, time, and location (this last form of data is often inferred or incomplete). These bits of data can be thought of as traces of human activity and can help explain how social media activity intersects with different facets of urban life. The data produced through social media is different from the traditional data formats. Social media data is unstructured and complex and often comes with noisy combinations of text, images, time, geolocations, and other metadata. Additionally, it is often too big to study through qualitative means or even to quantitatively analyze through normal spreadsheets. Methods used in this book include ways to summarize large amounts of data without having to read them individually, to understand community structures from networks of people, to measure sentiment and emotion from text and images, and to automatically cluster different types of images.

There are numerous social media platforms used extensively for a variety of reasons. Each of these platforms have unique functionalities and produce data in unique formats and access levels. The first and most important point is to differentiate between public and private social media data. Almost all social media platforms differentiate between public and private posts. Furthermore, each platform has unique policies governing the research access to data produced within their platform. This book solely focuses on public data that are publicly available through the platforms. Access to these datasets is through each platform's Application Programming Interface (APIs). APIs are the gateways to gain access to each of these platforms. Platforms such as Twitter, Reddit, and Telegram offer controlled means of collecting non-private data. Instagram has a very limited way of collecting data, and Facebook's API almost completely restricts data collection.

What one can learn from social media data is very much related to how the data is structured and how users use the platform. For example, the microblogging and multimedia platform Twitter, which may be the most influential platform and source of social media data among scholarly research. Users on Twitter publish tweets that contain up to 280 characters of text, and might or might not include images and videos. Within Twitter, users who "follow" another user, see their tweets within their "timeline". Twitter users can share an interest in others' tweets by "liking", propagate others' tweets by "retweeting", and reply to and quote others' tweets. Hashtags are used by users to identify themes and topics for their tweets and also get topics trending on social media. By clicking on hashtags, users can see other tweets that have utilized the same hashtags within their tweets.

Social media platforms, have rich and complex data ecosystems. Information, in the form of texts and images, propagates from users to users through retweets, quotes, replies, and hashtags. Each social media post collected from Twitter normally includes a text field but also comes with a time of posting and the source of the tweet. Researchers can use a combination of text, image, time, location, and sharing (social network) behavior of social media posts.

Datasets from social media are typically very big, often too big for humans to manually read and understand. Computational scientists have developed an arsenal of methods that aid in making sense of such large datasets collected from social media. These methods include learning from existing known datasets to make sense of new ones, finding patterns and groups in data, extracting meaningful information from images and text, and finding important actors in a social network.

Supervised Machine Learning

Sometimes datasets have embedded information that can be useful. For example, tweets may be labeled by the user or by a third party based on different positive or negative sentiments. The intuition behind supervised machine learning (ML) is that clever methods can be used to learn from our existing datasets, so there is no need to repeat the cumbersome process of labeling data when new datasets are available. In other words, an ML algorithm can be trained using existing labeled datasets to predict the same labels for new unlabeled datasets.[11] The potential of supervised ML is limitless; it can score tweets based on sentiment,[12] or emotions[13]; it can train algorithms to find objects in images[14]; it can predict a new word, or sentence based on previous words.[15] ML models are not perfect; it is almost impossible to find models that are always accurate. ML researchers have produced many models that can help extract information from datasets. Often, reports about the accuracy and efficacy of those models are published. Training robust supervised models and subsequent verification require large datasets and considerable efforts. In this book, existing ML models created and verified by the national and international research community are routinely used rather than creating new models. This creation of standard "toolboxes" is standard practice in the data analytic community.

Transfer Learning

Supervised ML algorithms are numerous ranging from very simple models to extremely complex. One way to explain this learning is that models can gain some knowledge about a particular type of dataset. They may have learned that certain words might come after others or that certain colors or objects in images tend to be close to each other. There are cases where this knowledge learned from a previous model can be used on a new dataset. The task could be training a new model using new labels but with a small dataset, or using the outputs of the model to group the data without having labels. For example, a model might be effective at finding objects in images such as cats, dogs, chairs, buildings, trees, and humans. This model has learned a combination of pixels and colors that resemble a cat, or a chair. For example, it might be useful to classify images of rural areas and urban areas. Using transfer learning (TL)[16], it is possible to take advantage of the knowledge learned from the more robust basic model of what buildings and trees look like, and then use that knowledge to label pictures with more trees and vegetation as rural, as opposed to one with lots of cars and buildings as urban. This process is called TL, and it is useful in some cases.

Unsupervised ML

There are many cases in which there are no actual labels in the data. However, it is useful to find patterns, clusters, groupings, or communities.

Unsupervised ML[17] includes algorithms that are developed for finding potentially meaningful patterns in data without having existing labels. For example, it would be useful to divide a large corpus of images from Reddit into multiple groups in order to find out what each of those "clusters" mean. During the clustering process semantic differences can create clusters that are meaningless to human interpreters. This process requires interpretation to identify potentially meaningful patterns or clusters that can help explain the meaning of a group of images. In addition to image clustering, this process can automatically group text documents into topics or group sources of tweets based on their usage of specific words.

Understanding Text

An important feature of social media posts is the text data of each social media post. Depending on the platform, the length (number of characters) of the document could vary. But the unstructured nature of text data requires methods that are different from the conventional structured tabular datasets.

When dealing with millions and millions of text documents, one of the first steps is to get a general sense of the patterns and topics in the data. Topic modeling is the term used for the process of statistically summarizing and categorizing large numbers of text documents into different topics without any supervision. Topics are defined as "distributions over a vocabulary of words that represent semantically interpretable 'themes'".[18] Topic models usually provide a series of words or phrases that statistically correlate with each other within a given corpus of documents. A good topic model can be interpreted as semantically coherent and meaningful.

Topic models work by first transforming text documents into numerical representations. One of the simplest ways text documents are converted to numerical vector representations for conducting statistical analysis and topic modeling. The order of words in a document is assumed to be unimportant. If a (very long) list of the unique words in all of the documents is compiled the frequency of occurrence for each word can be computed as a numerical vector representation of each document (the document-term matrix). The document-term matrix is a very large table with rows representing each document, columns representing each unique word in all of the documents, and the values as the number of times each word has occurred in that document. Refinements to these methods include identifying common words such as "the" or "is" (known as stop words) that in most documents carry very little meaning. Another refinement is to weigh the frequency of a word in one document by the number of documents that contain that word in all of the corpora, thus giving more importance to words that are more unique to each document. The name for this process is Term-Frequency Inverse-Document-Frequency vectorization which is used in many different topic modeling algorithms. There are many other more complex ways texts are transformed into vectors that are beyond the scope and interests of this book.

Different topic modeling algorithms use such representations to provide useful information. One of the most utilized topic models is called Latent Dirichlet Allocation (LDA).[19] Given a corpus of text documents and a pre-determined number of topics, LDA provides multiple forms of useful information. First, it provides a number of topics. Each topic is simply the collection of all the unique words in all documents ordered by how important those words are to each topic. For example, LDA can extract 20 topics from a collection of social media posts about food. Topic 1 might have the words Pizza, Pepperoni, Mushroom, and Topping as the most representative terms, and topic 2 might include Coffee, Latte, Espresso, and Milk as the most representative terms. Along with the words, LDA also outputs numbers that represent the probability that each word is important within the topic. All the other 18 topics will include a similar list of words that represent how important those words are for each topic.

LDA can also infer topics from a document, which means that after creating the mentioned list of 20 topics for our food-related social media posts, the algorithm can infer the most representative topics from a document. The topic distribution for each document would be a list of all 20 topics with the probability of that document belonging to each topic. For example, a hypothetical output from LDA for the tweet "This restaurant has really good pepperoni pizza, but the beef sandwiches were not really good" would be 50% Topic 1 (Pizza), 40% Topic 10 (sandwiches), and 10% divided between all other topics. Since there are no coffee-related words in this document, the probability of this document belonging to topic 2 (Coffee) would be close to zero.

Another Topic Modeling approach that is used in this book is Top2Vec.[20] Top2Vec uses TL or other embedding techniques to create vector representations of words, documents, and topics. It then uses clustering to find dense areas of documents that it calls topics. Since word vectors are in the same space, Top2Vec can show the most representative (closest) words to each topic, as well as most representative topics for each document, or the most representative topic to a word. This technique is very useful because you can search a large number of documents by a specific keyword and find different topics that are close to a word. For example, you might search for the word Coffee, and find multiple topics and documents that are related to this. Top2Vec has a very efficient python library[21] that makes conducting these tasks relatively simple.

LDA and Top2Vec provide a way to summarize and find patterns from a very large number of text documents. And, They also allow the identification of documents that might be related to specific extracted topics. Topic modeling is a very useful unsupervised method for making sense of a large corpus of text documents. Aside from LDA and Top2Vec, there are numerous other topic modeling methods (for example, the Structural Topic Model (STM))[22] that have an output but with some benefits or drawbacks that we will utilize throughout this book.

Many supervised methods exist for studying social media posts. Named Entity Recognition (NER)[23] covers a body of algorithms that help tag different words in a document. For example, it is very common to find mentions of places, persons, organizations, money, and dates from a text. NER is very useful for filtering and summarizing text documents. For example, a NER model can categorize a sentence such as "Chicago protesters will be gathering in Grant Park on January 4th" and label Chicago as a place, Grant Park as a place, and January 4th as a date. A combination of mentions of future dates and places can imply when a future event might occur, in this case, a protest in Grant Park on January 4th. However, NER models can be prone to inaccuracies, tagging words as organizations as persons or vice versa.

As mentioned in the supervised ML section, two popular classes of models are sentiment analysis and emotion analysis. Sentiment analysis gives a score to text based on a spectrum of positivity and negativity. For example, these models can score sentences like "Many people are dying due to COVID-19 and no one cares to do anything" as very negative, and "Vaccines are finally here, rejoice!" as very positive. Text emotion analysis is similar, but instead of scoring text documents based on positivity or negativity, it provides scores for different emotions, such as sadness, anger, joy, and disgust. Unsupervised models can be combined with sentiment analysis, to see if clusters of users have different sentiments towards a specific topic.

Understanding Images

From selfies, memes, landscape images, and even photographs that include text, there is a vast range of knowledge that can be extracted from images on social media. However, the process of understanding and analyzing images is part of the subdiscipline of computer vision that is very different from natural language processing and analysis of other forms of data. Images consist of pixels, each in a grayscale or color format. Colors are described using various methods such as RGB (red, green, and blue) and HSL (Hue, Saturation, and Lightness). While in natural language processing each word has a corresponding meaning, in computer vision each pixel is inherently meaningless. It is the combination of different pixels with specific colors that create visuals that are so simple for humans to understand. However, each set of pixels, even with the slightest difference, is a completely different set of pixels for a computer. In other words, a picture of a cat, and a slightly rotated picture of a cat, are indifferentiable for humans, but very different for computers.

Until the recent advancements in ML and artificial intelligence, computer vision consisted of clever algorithms that allowed for the detection of semantic information such as edges, blobs, text, and numbers from text. However, recent advancements in Artificial Neural Networks (ANNs)[24] allowed for a boom in how images are processed and analyzed by computers. In supervised ML algorithms, the aim is to start with an input dataset and learn complex patterns that would help with classifying each input into multiple classes. It is based on a set of connected units or nodes called artificial neurons, which loosely model the neurons in a biological brain. Each connection transmits a signal to other neurons. An artificial neuron that receives a signal processes it and then it can signal neurons connected to it. Neurons and edges typically have a weight that adjusts as they learn. The weight increases or decreases the strength of the signal at a connection. Neurons are collected into layers that perform transformations on their inputs. Signals travel from the first layer (the input layer) to the last layer (the output layer) after passing the layers multiple times. After proper evaluation of a model's accuracy and effectiveness, the label with the highest probability can be taken as the classification output of the model. In computer science, ANNs with a very large number of hidden layers are called Deep Learning algorithms.

The specific algorithm that is responsible for most advancements in computer vision is called Convolutional Neural Networks (CNNs).[25] The advantage of CNNs is that through multiple stages, they start by going over each pixel, and finding patterns in adjacent patterns. For example, in the first stage, a CNN will detect edges in a large dataset of cats versus dogs, in the next stage, it might start to find patterns in edges and detect shapes such as eyes and ears, and finally, arrive at a representation of a dog. It is important to note that the computer does not have a notion of dogs versus cats, rather it can differentiate between features in images that are labeled as being cats or dogs. There have been numerous very robust models developed, some can detect and find hundreds of objects, such as humans, chairs, buses, cars, and spoons, and others can detect emotions from pictures of faces.

An alternative strategy is TL. TL is a research method in ML that focuses on storing knowledge gained while solving one problem and applying it to a different but related problem. For example, knowledge gained while learning to recognize cars can be used to recognize trucks. Instead of getting class predictions from a new CNN model, numbers from the last hidden layer of a model can be used. These numerical representations (called embeddings) are very useful representations of the images. Extracted embeddings of images can be used to find images that are semantically similar to each other. For example, the input of an image of a place in Chicago can be used to find other pictures that include similar elements within them. Such computer vision methods are used several times in this book. In each occurrence, more details will be provided.

Understanding Social Networks

The word social, in social media, refers to how people engage in interactions with others through activities such as following, subscribing, retweeting, liking, and commenting on others' media posts. Studying these activities can highlight interesting social phenomena such as online communities, groups, and echo chambers. Furthermore, it can help us, for example, to find actors within these groups that are more influential or connected. A group of analytical methods that focus on studying such phenomena includes social networks or graphs.[26] Social networks are essentially a way of modeling and studying the world in terms of elements and how they are connected to each other. Such methods are already a part of our lives.

Examples range from web pages and how they hyperlink to each other and authors of scientific articles who reference the same articles in their publications, to networks of interactions between proteins in a cell and connections between neurons in our brains. In all of these examples, the elements (Webpages, Twitter users, proteins, neurons) are mostly thought of as nodes and their connections (hyperlinks, retweeting, interactions, and connections) are thought of as links or edges. By modeling such systems as networks, a wide array of algorithms and metrics developed by graph theorists and network scientists can be used. This book utilizes some of these methods for understanding the relationship between social media usage, the local communities, and urban spaces.

There are numerous ways social media data can be modeled as networks. A network of Twitter users who follow each other consists of users as nodes and the link between the nodes illustrates users following other users. Since following is directional, meaning that following on Twitter is not mutual, the links in such networks are directional and include two links between two nodes denoting the direction of following. You can imagine that there are far more links directed towards the account of the president of the US as there are links directed from the account to others. Such networks are called directed networks. To analyze the network of users of a specific hashtag on Twitter, a link is created between two accounts that have used specific hashtags between themselves. Since accounts might share more than one hashtag, such networks are often weighted, meaning that a number is assigned to the links between two nodes based on the number of hashtags that are shared between them. Using weights, it is possible to identify the users who have a stronger topical connection and ones that have a weak connection. The concepts of directed and weighted networks are used in many of the case studies that follow this chapter.

Viewing a phenomenon as a network provides a wide variety of metrics, properties, and methods to analyze and understand that phenomenon. In the Twitter example, the number of links incoming or outgoing from a node is a very simple but a powerful property of that node. An account on Twitter that has many followers but follows few accounts is a node that has many incoming links, but very few outgoing links (an account typical for celebrities). On the other hand, some accounts follow a large number of other accounts but have very few followers typical of a Twitter bot that automatically follows accounts. In network science, the number of links of a node is often dubbed as the "degree" of that node, and consequently, the number of incoming links is called "in-degree" and the number of outgoing links is called "out-degree".

Another interesting concept in a network analysis is the concept of a path between two nodes. A path is a set of links that connects one node to another. This concept is naturally understood in a street network with intersections as nodes and streets as links. A path between two intersections A and E is a collection of streets that connects those intersections. One path would possibly connect intersection A to B, B to Z, and Z to E. So, one path from intersection A to E would go through links connecting nodes B and Z. Obviously, there are many paths between two intersections in a city as there are many ways of getting from point A to point E. The shortest path between two nodes is the selection of paths with the least distance. So, what is the distance between these two nodes? In physical networks such as streets, the concept of distance is easy to understand. The shortest path is the path with the least physical distance. In other networks, the concept of distance might be viewed differently. In our follower network, the path between two nodes could simply be the number of links that connect them. For example, if you do not follow Kanye West, but you follow a friend who does, the distance between you and Kanye could be two links. If your friend was a famous journalist, Kanye followed your friend, and your friend followed you, then the distance between Kanye and you would also be two. Otherwise, the distance between Kanye and you could be much longer, perhaps six.

Networks are used ubiquitously through our GPS map applications. Those apps use networks to efficiently find the shortest paths in a street network and draw those directions for us. The shortest paths can also be used to derive a metric for the relative importance of nodes in a network. In network science, another term used for the importance of a node or link is "centrality". If people always drive the shortest paths between their origins and destinations in a street network, the streets (links) with the highest number of shortest paths between them have the most amount of traffic. The number of shortest paths that go through a node or link from all nodes to all nodes of a network is called Betweenness Centrality. These centrality concepts often have different meanings and implications in different networks. For example, in a network of users who retweet each other, a user with high betweenness centrality might be a popular news aggregating account. That account may be a hub or bridge. If suddenly, that account was blocked by Twitter, many users might be cut off from some of the information in the network.

Another related concept of centrality based on shortest paths is Closeness Centrality. If we calculate the shortest paths from a node to all other nodes and take the average of those distances, we arrive at a metric of how far that node is from others. By inverting that number, a metric can be calculated of how close a node is to all other nodes. In a follower network, if an account has a very large number of followers, it is likely to have a high closeness centrality and their tweets might propagate faster in the network. If a user has a low number of followers, it does not necessarily mean that they have a low closeness centrality. In such a network, highly influential friends and neighbors are important indicators of a node's centrality. If a user has a follower with a high closeness (potentially high number of followers) that users' closeness would also be high. In the example of friendship with Kanye West, if Kanye followed another account and retweeted one of the tweets, it is likely to reach many people, even though that account might have a few other followers. This concept was famously an intuition for Google's founders to develop an algorithm called PageRank[27] that was used within the search engine.

In networks, often nodes are not all connected to each other. Some sets of nodes might be highly connected to a group but not others. In that sense a group of nodes might form a "community" of nodes that are highly and densely connected to each other but not others. In social media, gamers would follow a set of accounts and form a community that is not highly connected to a community of academics. Obviously these communities are not exclusive since we might likely have individuals who are gamers and academics and may well be members of both communities. These subtleties in understanding communities have led to an array of methods in "community detection" that, with certain assumptions, allows finding and extracting communities in networks.

Obviously, there are many other important concepts in network science to be discussed that are beyond the scope of this book. However, since in certain chapters of this book methods derived from network science are used, hopefully this introduction can help with understanding and inspiring some of the research.[28]

Spatial and Temporal Analysis

Even though this book focuses on cities, it does not have a heavy focus on geospatial analysis methods. This is in part because geospatial analysis requires geospatial data and social media data do not always come with reliable spatial information. A very small portion of tweets or Instagram posts is geotagged. Geotagged posts are accompanied by specific latitude and longitude information that can be used for mapping. At various parts of this book, when appropriate, such information is visualized using Geographic Information Systems. Even though the majority of social media posts are not geotagged, it does not mean that they do not contain geographic information. In many cases, the text body of posts includes mentions of specific places, cities, or countries. Using NER, it is possible to extract such information from the text. Such extracted place names can then be fed into geocoding software that would translate a text into a location. The geocoding process is essentially the one that is used when you search a name of a restaurant on Yelp and see a point on the map. In this book, in some cases, we use this process to get more geospatial data from non-geospatial tweets. However, since this is a cumbersome and expensive process, we limited doing this to cases where we absolutely needed to.

Notes

1 Zeisel, John. (2006). *Inquiry by Design: Environment/Behavior/Neuroscience in Architecture, Interiors, Landscape, and Planning.* New York: W.W. Norton & Company.

2 Chapman, Lee et al. (2018). "Investigating the emotional responses of individuals to urban green space using twitter data: A critical comparison of three different methods of sentiment analysis." *Urban Planning, 3*(1), 21–33.

3 Collins, Craig, Samiul Hasan, & Satish V. Ukkusuri. (2013). "A novel transit rider satisfaction metric: Rider sentiments measured from online social media data." *Journal of Public Transportation, 16*(2), 2.

4 Glaser, Barney & Anselm L. Strauss. (1967). *The Discovery of Grounded Theory.* Chicago, IL: Aldine.

5 Page 7. Charmaz, Kathy. (2006). *Constructing Grounded Theory: A Practical Guide through Qualitative Analysis.* London: SAGE.

6 Page 667. Charmaz, Kathy. (2001). "Qualitative interviewing and grounded theory analysis." In Jaber F. Gubrium & James A. Holstein (Eds.), *Handbook of Interview Research.* Thousand Oaks, CA: SAGE.

7 Charmaz, Kathy. (2006). *Constructing Grounded Theory: A Practical Guide through Qualitative Analysis.* London: SAGE.

8 Page 79. Braun, Virginia & Victoria Clarke. (2006). "Using thematic analysis in psychology." *Qualitative Research in Psychology, 3*, 77–101.

9 Page 145. Creswell, John W. (2009). *Research Design: Qualitative, Qualitative, and Mixed Methods Approaches.* 3rd edition. Los Angeles: SAGE.

10 Hine, Christine. (2000). *Virtual Ethnography.* London: Sage.

11 For a survey of Supervised Machine Learning Methods also called classification, please refer to Kotsiantis, Sotiris B., I. Zaharakis, & P. Pintelas. (2007). "Supervised machine learning: A review of classification techniques." *Emerging Artificial Intelligence Applications in Computer Engineering, 160*(1), 3–24.

12 Kouloumpis, Efthymios, Theresa Wilson, & Johanna Moore. (July 2011). "Twitter sentiment analysis: The good the bad and the omg!" In *Proceedings of the International AAAI Conference on Web and Social Media* (Vol. 5, No. 1). Palo Alto, CA: AAAI

13 Wang, Wenbo, Lu Chen, Krishnaprasad Thirunarayan, & Amit P. Sheth. (September 2012). "Harnessing twitter 'big data' for automatic emotion identification." In *2012 International Conference on Privacy, Security, Risk and Trust and 2012 International Conference on Social Computing* (pp. 587–592). New York City: IEEE.

14 Zhao, Zhong-Qiu, Peng Zheng, Shou-tao Xu, & Xindong Wu. (2019). "Object detection with deep learning: A review." *IEEE Transactions on Neural Networks and Learning Systems, 30*(11), 3212–3232.

15 Brown, Tom B., Benjamin Mann, Nick Ryder, Melanie Subbiah, Jared Kaplan, Prafulla Dhariwal, … & D. Amodei. (2020). "Language models are few-shot learners." *arXiv preprint arXiv:2005.14165.*

16 Pan, Sinno Jialin & Qiang Yang. (2009). "A survey on transfer learning." *IEEE Transactions on Knowledge and Data Engineering, 22*(10), 1345–1359.

17 Xu, Rui & Donald Wunsch. (2005). "Survey of clustering algorithms." *IEEE Transactions on Neural Networks, 16*(3), 645–678.

18 Roberts, Margaret E. et al. (2014). "Structural topic models for open-ended survey responses." *American Journal of Political Science, 58*(4), 1064–1082.

19 Blei, David M., Andrew Y. Ng, & Michael I. Jordan. (2003). "Latent dirichlet allocation." *The Journal of Machine Learning Research, 3*, 993–1022.

20 Angelov, Dimo. (2020). "Top2Vec: Distributed representations of topics." *arXiv preprint arXiv:2008.09470.*

21 https://github.com/ddangelov/Top2Vec.

22 Roberts, Margaret E. et al. (2014). "Structural topic models for open-ended survey responses." *American Journal of Political Science, 58*(4), 1064–1082.
23 Nadeau, David & Satoshi Sekine. (2007). "A survey of named entity recognition and classification." *Lingvisticae Investigationes, 30*(1), 3–26.
24 Nielsen, Michael A. (2015). *Neural Networks and Deep Learning* (Vol. 25). San Francisco, CA: Determination Press.
25 Gu, Jiuxiang, Zhenhua Wang, Jason Kuen, Lianyang Ma, Amir Shahroudy, Bing Shuai, ... & Tsuhan Chen. (2018). "Recent advances in convolutional neural networks." *Pattern Recognition, 77*, 354–377.
26 Barabási, A. L. (2013). "Network science." *Philosophical Transactions of the Royal Society A: Mathematical, Physical and Engineering Sciences, 371*(1987), 20120375.
27 Page, L., S. Brin, R. Motwani, & T. Winograd. (1999). *The PageRank Citation Ranking: Bringing Order to the Web.* Stanford InfoLab.
28 Refer to books by Barabasi: Barabási, A. L. (2013). "Network science." *Philosophical Transactions of the Royal Society A: Mathematical, Physical and Engineering Sciences, 371*(1987), 20120375. or Newman: Newman, M. (2018). *Networks.* Oxford University Press. for a complete introduction to networks.

PART 1
Protests

3

BLACK LIVES MATTER[1]

As this book was being written in 2020, Black Lives Matter protests had become frequent, insistent reminders of the racial legacy in the United States. These protests have a great deal in common with social-media-influenced worldwide protests in the last ten years, such as in Spain, Egypt, and Hong Kong. This chapter presents a detailed analysis of one particular Black Lives Matter protest in Charlotte, North Carolina.

On Tuesday, September 20, 2016, Keith Lamont Scott was shot and killed by police officer Brentley Vinson. Spontaneous protests associated with the Black Lives Matter movement took place in the aftermath of the shooting for three nights. National and international news media reported the events.

Scott was in his car parked next to Vinson, an undercover police officer who was at the location on an unrelated investigation.[2] Additional officers were requested and when they arrived, Scott was ordered to get out of the car by the police. He did not immediately respond but did eventually exit the car and was standing with his arms at his side. Four shots were fired. There is a difference of opinion between the police who believed Scott was holding a gun and the family who believe he was holding a book. Rakeyia Scott, Keith Lamont Scott's wife, broadcast part of the incident via Facetime live that lasted for 2 minutes and 12 seconds. In this broadcast, she pleads with the police not to shoot her husband, tells the police he has TBI (traumatic brain injury), and pleads with her husband to remain still when he exits the car. The first gunshot is heard in the video but the camera is pointing at Rakeyia rather than the shooting.

Protests on the night of the shooting[3] began on Old Concord Road near the home of the Scotts, following the dissemination of the news on social media. The protest began at dusk. Later the same evening, the protest moved from Old Concord Road to the intersection of a major freeway (Interstate 85) near the original site. All four lanes of the traffic were closed to traffic.

DOI: 10.4324/9781003026068-4

Minor property damage occurred during the protests, and there were minor injuries to several police officers. One person was arrested.

On the night of September 21, organized protests were located at center city Charlotte. The protest started at Marshall Park on the fringe of the center city. The protest later moved into the corner of Trade and Tryon, the center of the city. The police attempted to control the protests using tear gas and organized police lines to force them out of the center toward the Epicenter entertainment complex. Forty-four people were arrested and nine protesters were injured. Two officers received minor injuries. The vandalism occurred in several blocks of the center city. One protester was shot in the head and later died. Police arrested a suspect who confessed to the murder.[4]

On the night of September 22, a citywide curfew was imposed from midnight to 6 am. Protests were peaceful and orderly. The protestors blocked interstate I277 in the center city for a brief period before the police, State Highway Patrol and the National Guard, forced them to disperse without major incident.

Cities are the place where since 2013 the majority of the world's population live, work, and innovate, but they are also the places where people face grave injustices and unfairness. The American city has consistently been a place where racial dynamics and inequality come to light. The civil rights movement in the 1960s took to the streets and focused on the public domain in an effort to narrow the racial divide. However, residential segregation has consistently remained a major feature of many American cities.[5] These inherent racial biases are evident in many other aspects of cities and societies in the United States. African American neighborhoods are prone to food disparities,[6] higher disability rates,[7] and an unfair education system.[8]

Social media has changed the way activists try to address these injustices and increase the participation of underrepresented populations.[9] The Black Lives Matter movement has spurred demonstrations against ongoing police brutality, racial segregation, and injustice in the United States. Black Lives Matter is expressed using social media with the iconic #BlackLivesMatter hashtag, facilitating urban protests in many cities in response to the deaths of unarmed African-Americans by police.[10] One of the key features of the Black Lives Matter movement, as highlighted by the hashtag, is the integral role played by social media. The role of social media worldwide can be seen in the urban protests and revolutions such as the Arab Spring in Egypt and Tunisia, the Green movement in Iran, and the Indignados movement in Spain.[11] Black Lives Matter supporters not only use social media as a means to show individual support, but they also use this tool to organize protests, connect with other supporters, communicate and discuss their goals and demands, and transmit their agenda and their peaceful message.[12]

The focus of this chapter is on this instance of a Black Lives Matter protest that occurred in Charlotte as a response to the fatal shooting of Keith Lamont Scott. The study adopted a mixed method including social media, the local community, and urban space to capture this heterogeneous data (Illustration 3.1). To study the protesters' perspectives on the significance of

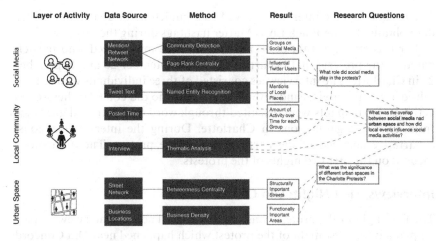

ILLUSTRATION 3.1 Mixed method research approach.

urban spaces in their activities, a series of interviews was conducted with activists in Charlotte. To study the protesters' use of social media, a large corpus of tweets was analyzed that included keywords related to the Charlotte protests. The analysis included natural language processing and analysis of social network mentions and retweets on Twitter. Urban space was also studied as a factor in the generation of social media activity.

Two questions concerning location were important in this case. First, did the location of members of online communities correlate with interests and attitudes of the community; put more simply, do local people differ in a significant way from non-locals (national or international). Second, does the intensity of social media communication significantly impact "events on the ground", or put another way does physical location generate social media activity?

Local Community

For the qualitative analysis of the local community, interviews were conducted with protesters who were among the first to participate. The sessions were semi-structured, with some open-ended questions. The interviews mainly consisted of discussing the multiple days of protest in Charlotte while focusing on the "why" and "how" of activists' actions. Questions regarding the locations of protests and the significance of those places were asked. The different types of social media they used and the role each social media played in their actions was also a focus. The information gathered from the interviews led to an understanding of the strategies adopted by protesters in both social media and urban space,

> how protesters reach decisions to take actions to make demands on social media, and
> how urban space was understood and selected for protest actions.

The results of these interviews served as a guideline for analyzing Twitter data obtained using Black Lives Matter hashtags during the protests.

For this study, a total of 10 individuals were interviewed who participated in the protests that occurred between September 20 and September 23 in Charlotte. The interviewees consisted of three individuals who did not self-identify as activists and seven individuals who did consider themselves activists. Participants were recruited through word of mouth and guidance from activist organizations in Charlotte. During the interview sessions, a narrative emerged of the participation in the protest. The discussions focused on the first two nights of the protests.

Interviews, First Night: Old Concord Road

The first night of the protests was unplanned. All activist interviewees participated in the first night of the protest which happened near Old Concord Road and Harris Boulevard in Charlotte. The initial communication was organized through Facebook chat, and organizing efforts were coordinated via text messages to rally protesters to the shooting location. Based on the observations of the interviewees, physical proximity to the location of the incident was the first draw for the initial student group. One of the interviewees who did not identify as being an activist mentioned that he first heard people shouting and went to the location to observe. It was only after hearing about the details of the event from the other protester that he started to protest against the police.

When asked why they went to Old Concord Road, almost everyone mentioned the desire to give support to the family of the victim. Additionally, it was the most natural place to go to protest as it was close to UNCC and many of the protesters were students. When asked about the implications of choosing Old Concord Road, one of the activists described how the physical nature of Old Concord Road impacted the protesters: "The reality is that it [the protest] had to be big because that area is so separated and it would be easy for police to surround or block off people". Later the same evening, the protest moved from Old Concord Road to the intersection of a major freeway (Interstate 85) near the original site. The fact that Old Concord Road is not well-connected (which made it easy for the protesters to be contained) led to this tactical shift in the location in an effort to make the message louder.

Regarding how the first-night protest grew in size and how people were motivated to participate, interviewees replied that Facebook live, Periscope, and Instagram were all used to livestream the events. One of the interviewees said, "Everyone was live streaming, so when they got one of our phones, others would still show what was happening". The news coverage, along with texting friends and live videos helped the first night of the protest to become loud and impactful. "By the end of the night, there were people protesting who came from Raleigh (the capital of North Carolina). A lot of them stayed with me that night".

Interviews, Second Night: Uptown Charlotte

The second night of the protests happened with more organization and planning. The location of the protest changed from the University Area to Uptown Charlotte. All of the interviewees told a similar story about the sequence of events. One interviewee who did not identify as being an activist said that she was contacted by one of her friends and was told about a gathering at a church in Uptown. The church was where some of the in-person organizations took place. One of the activist interviewees was also present in the church.

Most of the other interviewees started the second night's protest in Marshall Park, a park near an important government building in Uptown Charlotte without residential uses around it. Marshall Park is a place where most protests in Charlotte are sanctioned: "Marshall Park is one of the most central locations in the city, it's where all protests pretty much start at. It's also right by the jail, government center, etc.". Another interviewee said, "It's easiest to find parking for free around Marshall Park and it's easy to identify and use as a meeting place for people coming from different directions". Another interviewee had a different perspective about the park: "Even though it's close to Uptown, it is still separate from most of the activities…".

After Marshall Park, protesters flowed into the center of uptown, namely the intersection of Trade and Tryon streets which one of the activists described as "literally the center of the city". Another interviewee described this interaction as "I think that Trade and Tryon ended up being an important place because police tried to corner people into that area". This shift caused a direct response from police officers who were in fact not concerned with the protests in the sanctioned area of Marshall Park. Epicentre, a mixed-use entertainment, retail and residential project, is another location in the immediate vicinity of Trade and Tryon where many of the protests were pushed: "Epicentre and the areas surrounding represents really what this city was made for right [wing], white people. It represents that these are the areas to disrupt".

There were other places that were mentioned by some of the interviews but were not as prevalent. For example, a football stadium is also very close to Trade and Tryon:

> It was really difficult to disrupt people in front of Panther's games at the stadium because the stadium is technically "private" property so in order to avoid arrest we had to stay on "public" property which made it more difficult to interrupt as we were separated from the stadium and the people whose attention we were trying to get.

Social media played an important role in motivating people to participate in the protests. One of the protesters who did identify as being an activist said that he would not have participated if he did not see a livestream of the protest in Trade and Tryon. "When I saw the video on Facebook, I just had to go…. you could see exactly what was happening and that is powerful".

Encrypted text messaging was also considered an important tool for communication. Activist interviewees mentioned that after the first night of protests many phones were investigated by the police. In response, protesters started using encrypted text messages for security purposes. Networks of friends were also a very important factor in bringing more people into the protest as described by one of the interviewees: "if my roommate didn't text me, I wouldn't have gone because I had a long day and I was really tired".

Throughout the protests, different types of social media were used but for different purposes. Events on Facebook were used to organize people for future protests. Photo sharing apps such as Snapchat and Instagram were used to share photos of protests as they happened. Chat applications were used to give support to other people and ensure their safety. Other social media such as Twitter were used to read and publish news about the protest in real time. Finally, livestreaming the event played a crucial role in motivating people and contextualizing the protests for people who were not present at the protests.

Social Media Analysis

In order to analyze social media related to the Charlotte Protest through social media, Twitter GNIP API was used to obtain a dataset of tweets that included keywords and hashtags such as #keithlamontscott, #charlotteprotest, and #charlotteriots. GNIP API allows for access to a complete collection of tweets based on the requested query. The dataset consists of approximately 1.3 million tweets between September 20 and 23.[13] In this section, the temporal and spatial nature of these tweets are studied

The analysis of social media data consists of two main parts. First natural language processing was used to extract meaningful information regarding spatial data from tweets, using Stanford Natural Language Processing (NLP)[14] NER to extract location information from the tweets (Manning et al., 2014). NER is a method that tags words in a document on the basis of location, date, person, organization, or other similar tags. It afforded detection of mentions of Charlotte places and other spatial information from the texts of the tweets.

Second, social network analysis focused on user mentions and retweets to understand the different actors and their roles in social media. In this network, Twitter users are nodes and retweets or mentions are edges. For example, if user A retweets user B, there will be a directed edge between B and A. If this retweet act happens multiple times between these users, the weight of the edge between them will increase. This method results in a weighted directed social network.

Different social network analysis methods and metrics were used to understand the features of this network and how different groups of people interact with each other on Twitter in the context of a protest. The first method to analyze the social network is community detection which automatically categorizes the nodes into different tightly connected groups.

Community detection allows us to understand whether different users shape different clusters and whether they behave differently in these clusters. Maximum modularity community detection is used to automatically detect major community structures in the social network.[15]

The Pagerank algorithm is used to analyze the importance of Twitter users in the network.[16] Pagerank was originally created by Larry Page of Google to measure the importance of Web pages in the World Wide Web based on their number of links and their proximity to other important nodes. The PageRank algorithm produces a probability distribution for each node representing the likelihood of a random surfer ending up on a specific node.

The extracted information from tweets, as well as the time of posting, and the location of posting (if available), can then be compared with the results of the interviews from our two case studies.

The final part of the analysis consists of merging the results of the social media data analysis and interviews through the context of urban space. First, an analysis of users' use of detailed spatial information in their tweets was calculated and serves as another measure of "importance" for nodes in the networks by simply counting the number of times nodes are mentioning specific places in their tweets and studied nodes with high place mentions in the context of their social network.

Spatial Information in the Charlotte Protests Tweets

The locations mentioned by our interviewees were compared to their surrounding areas to see whether they stand out based on different demographic and geographic information. Illustration 3.2 shows that the areas around Uptown Charlotte are among the highest in terms of density of businesses and street connectivity. However, that was not the case for the area where the first night's protests occurred. These results confirm the narratives gathered from protesters about the significance of these different areas. More specifically, the location of protests on the first night is significant since it began in the immediate vicinity of the shooting, that is, the victims' house. Later that night the protesters moved to a functionally important interstate highway. On the second and third nights, the protesters moved to more functionally or symbolically relevant locations near the crowded and commercial areas in the city. Illustration 3.2 shows maps for each of the mentioned spatial analyses. The darkest color indicates the highest business density and white the lowest; the thickest streets indicate the highest connectivity.

Extracting geospatial information is crucial to understanding the spatial nature of social media usage. Unfortunately, the tweets dataset is almost without geolocated tweets (less than 200 tweets have geolocation information). This lack of geolocation information is in line with the low rates of geolocation information on Twitter (around 1%). It could also be a strategic action to shield one's location.

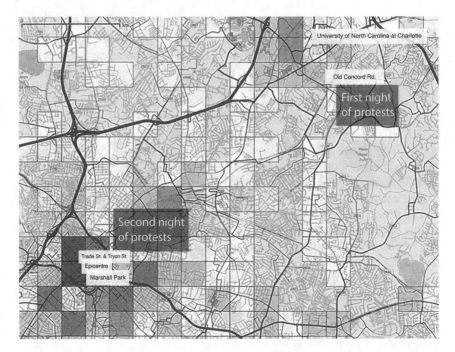

ILLUSTRATION 3.2 Business density and connectivity.

To rectify this problem and mine geolocation information from other sources, NER was used on all the tweet texts to find mentions of places in the text body of the tweets. The results of this NER analysis were saved in a database for further analysis. Illustration 3.3 shows a visualization of the count of each place that is mentioned in the dataset, with the size corresponding to the frequency.

The highest number of place mentions found in the dataset was Uptown Charlotte which fits well with the focus group results. Marshall Park, Trade Street, and Epicentre were also mentioned, both in the interviews and in the NER results. These results show that the tweets correspond to specific places where protests occur. Important places mentioned in the interviews are also found with higher numbers in the tweets' datasets. However, not all tweets contain mentions of specific places. In fact, the majority of the dataset contained tweets discussing the shooting or details of the subsequent events such as the request of the victim's family demanding the release of the video, comments on the video itself, or questions as to whether the victim held a gun in his hand or a book. Understanding how these places are mentioned and by which groups of users, helps us better understand the relationship between social media and public space. To do so, the results of the NER analysis are combined with social network community detection to understand how different groups of people use geographic information in social media.

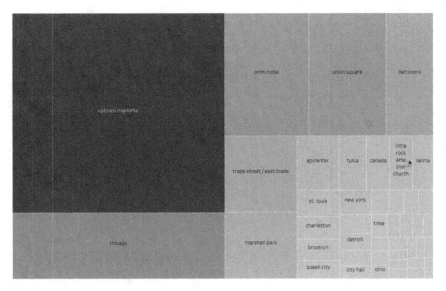

ILLUSTRATION 3.3   Tree map of specific place names.

Mention-retweet Social Network

A social network was constructed of the Charlotte protest by considering each Twitter user as a node in the network and the existence of a mention or a retweet between them as an edge. The constructed network was a directed graph with 341066 nodes and 899237 edges. The weight of edges in this network is the count of retweets or mentions between these two nodes. Each node also has the number of times they mention any of the detailed places in Charlotte such as Marshall park, Epicentre, Trade and Tryon, and Omni Hotel in the text of their tweets (the placeMention of each node).

Community detection was conducted on the whole network. As seen in Illustration 3.4 the results show two main clusters that are highly connected within the cluster and less connected to the other. The community detection produced other clusters which were too small to be visible. The PageRank of each node in the network was calculated, as a metric for the importance of each node in the network.

The results from these algorithms allowed exploration of the content of tweets and an understanding of the different groups of users and the different ways Twitter content was used in the Charlotte protest. To do so, the top nodes in the sympathetic and unsympathetic communities were examined based on their PageRank and placeMention. Tweets by the top 20 users in each group were read, along with their Twitter page (if it existed), and were tagged based on whether they were local to Charlotte, whether they were pro, neutral, or against Black Lives Matter protests. The occupation of the user was noted based on their Twitter user profile, and whether it was the Twitter handle of a person, an organization, or a news agency.

ILLUSTRATION 3.4 Social networks.

Studying the top 20 nodes in the network immediately highlights important distinctions within the Twitter data as a social network (See Illustration 3.5). First, within the top 20 most influential nodes in the network, some users are not local to Charlotte. Some users are against the protests and criticize the actions, and then there are users who are supportive of the protests. Moreover, users who are against the protests are from the "unsympathetic" community (colored in the tables and the graph visualization) and the people who are neutral and for the protests are almost uniformly from the "sympathetic" community (shaded in the tables and the graph visualization). These results show distinctly different approaches to the Black Lives Matter protest from different groups.

None of the top influential nodes in the unsympathetic community are local to Charlotte and there is an even distribution of local and non-local users in the sympathetic community (Illustration 3.5). This indicates that supporters or criticizers of these protests are more likely to retweet each other and belong to the same community. There is very minimal mention of detailed Charlotte locations in the unsympathetic community, and more mention of places in the sympathetic community.

To investigate more deeply the role place plays, the nodes with the highest mentions of detailed Charlotte locations were studied (Illustration 3.6). Except for one user, all the others with the highest mention of specific Charlotte places were either pro or were neutral to the protests and they were from the sympathetic community. Reading tweets from specific users shows that the users were not local, but in fact from the United Kingdom. All of

				Community 0: Mostly Against Black Lives Matter				
Id	Community	pagerank	placeMentions	Local	Position	Notes	Type	degree
142117	0	0.006485	1	No	Against	Media Analyst, Youtuber	Person	4161
49814	0	0.005166	0	Unknown	Against	Suspended/Removed Account	Unknown	24277
239813	0	0.004543	0	No	Against	News Agency	News Agency	2289
327629	0	0.004412	0	No	Against	None	Institution/organization	736
274031	0	0.003775	0	No	Against	Podcast Host	Person	2473
21218	0	0.003667	1	No	Against	Youtuber	Person	12239
321568	0	0.003598	0	No	Against	Journalist	Person	1404
141731	0	0.003381	0	Unknown	Against	Suspended/Removed Account	Unknown	2142
125980	0	0.003296	0	No	Against	Filmmaker	person	4073
245463	0	0.002838	1	No	Against	Twitter News	Institution/organization	1462
209889	0	0.002783	0	No	Pro	Prominent Politician	Person	3942
334553	0	0.002531	0	No	Neutral	Christian Conservative	Person	927
128052	0	0.002486	0	No	Against	Twitter Personality	Person	2896
332142	0	0.002230	0	No	Against	Twitter Personality	Person	1025
156594	0	0.002173	0	No	Against	None	Institution/organization	1224
1502.	0	0.002130	0	No	Against	Twitter Personality	Person	2476
246906	0	0.001975	1	No	Against	Twitter Personality	Person	1536
78340	0	0.001883	0	No	Neutral	Reporter	person	7425
165716	0	0.001856	2	No	Against	Twitter Personality	Person	286
63912	0	0.001762	0	No	Against	Suspended/Removed Account	Unknown	1535
44539	0	0.001716	0	No	Against	None	Institution/organization	6195

				Community 52: Mostly pro or neutral to Black Lives Matter				
id	Community	Pagerank	placeMentions	Local	Position	Notes	Type	degree
159764	52	0.015189	9	Yes	Neutral	News Reporter	Person	1953
341063	52	0.015010	0	No	Pro	Political Analyst and Reporter	Person	4241
334131	52	0.011286	1	No	Pro	Activist	Person	7170
155828	52	0.006851	2	No	Pro	Political analyst and activist	Person	5083
78783	52	0.005200	31	Yes	Neutral	None	News Agency	2193
286112	52	0.005058	0	Yes	Unkown	Charlotte police department	Institution/organization	12470
334280	52	0.004109	0	Yes	Pro	Journalist	Person	7279
134055	52	0.004095	15	Yes	Pro	Reporter	Person	1920
336391	52	0.004080	0	Yes	Neutral	Reporter	Person	987
151486	52	0.003871	0	No	Pro	Reporter	Person	3952
199473	52	0.003827	0	No	Pro	Journalist	Person	46
155816	52	0.003814	2	Yes	Neutral	Digital Reporter	Person	421
153098	52	0.003633	12	Yes	Neutral	None	News Agency	3136
336094	52	0.003466	1	Unknown	Pro	Twitter personality	Person	251
170847	52	0.003376	0	No	Pro	Activist	Person	6526
149381	52	0.003297	1	Unknown	Pro	News for the new America	News Agency	6033
340928	52	0.003250	2	Unknown	Pro	Activist, twitter personality	person	190
164509	52	0.002968	3	Yes	Neutral	News Anchor	Person	35
78036	52	0.002769	0	No	Pro	Activist	person	1091
165367	52	0.002728	1	No	Pro	Activist Protestor	Person	1350

ILLUSTRATION 3.5 Comparison of two major communities in the Charlotte protests network.

the 29 mentions were retweets of one tweet that included the word "Marshall Park". Furthermore, 12 out of the top 20 users with high placeMentions are local to Charlotte. Out of the top 100 users with placeMention (lowest count of five placeMentions), 85 nodes belong to sympathetic communities that are more likely to be supporters of Black Lives Matter or held a neutral stance.

Further investigation of this behavior is possible by comparing the temporal behavior of these two communities with the unfolding of events in the Charlotte protests. Illustration 3.7 shows social media activity in Twitter correlated with major events over the two-day period. The relationship of Twitter activity with protests in urban space is clear. Further, the temporal behavior of these two communities is indeed different. The "sympathetic" community has a first spike in activity that coincides with the first night of

id	Community	pagerank	placeMentions	Local	Position	Notes	Type	degree
303903	52	0.001482	56	Yes	Neutral	None	News Agency	3099
78783	52	0.005200	31	Yes	Neutral	None	News Agency	2193
123051	52	0.000464	30	Yes	Neutral	None	Person	912
25731	0	0.000000	29	No	Against	self employed	Person	13
190314	52	0.000198	23	Yes	Neutral	Blogger	Person	422
284497	52	0.000079	23	Yes	Neutral	None	News Agency	183
307904	52	0.000251	21	Yes	Pro	Journalist	Person	220
53720	52	0.000940	20	Unknown	Pro	Activist	Person	639
49516	52	0.000190	19	Unknown	Pro	Activist	person	249
226442	52	0.000124	19	Yes	Pro	artist, author, music	person	39
86847	52	0.000054	18	No	Pro	Musician	Person	140
224676	52	0.000685	17	Unknown	Pro	Unknown	person	325
111033	52	0.000069	17	Yes	Pro	Activist organizer	person	36
117766	52	0.000065	17	Yes	Neutral	None	News Agency	65
283555	52	0.000058	17	No	Neutral	News anchor	Person	195
134055	52	0.004095	15	Yes	Pro	Reporter	Person	1920
245421	52	0.000301	15	Yes	Pro	Activist	person	47
150045	52	0.000222	15	No	Pro	Activist	Person	110
117069	52	0.000071	14	Unknown	Pro	Don't know	person	51
125709	52	0.000975	13	No	Pro	Activist	Person	1020

ILLUSTRATION 3.6 Top users mentioning Charlotte places.

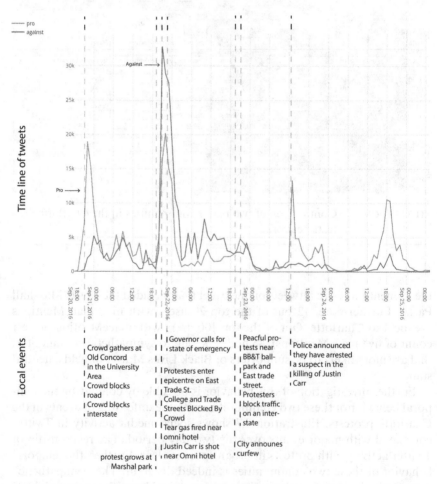

ILLUSTRATION 3.7 Social media activity during protest.

protests on Old Concord Road. This community has a spike of similar magnitude during the second night of protests, and two slightly smaller spikes during later events. The unsympathetic community has a single spike in activity on the second night of the protests corresponding to major national and international news coverage of these events.

Protest and Social Media

Recently, new terms have surfaced to describe an emerging form of activism. "Horizontalism" was coined by Juris and was used to describe how digital media has shifted the form of social movements to "leaderless" and "horizontal" movements.[17] Hardt and Negri use "swarms" as an analogy for describing how the new forms of communication create a collective intelligence that is more than the sum of the individual agents.[18] Others have criticized the idea of horizontal movements by emphasizing that it is in fact the "collective identity" created through the rapid sharing of ideas and symbols and is the main reason behind the integral role of social media in protests.[19]

The nature of social media and its effects on social movements are heavily debated. Alterman argues that even though social media such as Facebook and Twitter has in fact played an important role in mobilizing protesters in Egypt, news agencies such as Al Jazeera made a greater impact on the events of the Egyptian revolution.[20] In contrast, Earl and colleagues researched on a corpus of tweets related to the protest surrounding the G20 meeting in Pittsburgh in September 2009. They test a series of hypotheses regarding the usage of Twitter for sharing of protest locations, as well as the police action and conclude that Twitter has played an integral role as the primary form of organization in these protests.[21]

Although the conclusion set forth by Earl has been widely accepted, there hasn't been as much focus on the relationships and dynamics of place, activism, and social media. Hardt and Negri argue that in the current globalizing world, the place is not of primary importance. "The multitude", which they define as a new form of social class, is created in the globalizing world and by the new forms of network communication. They claim that "the multitude" is irreducible to its individual agents and does not have a place.[22] Gerbaudo, however, criticizes the idea and argues that the place which is occupied by activists highlights a form of unity and togetherness that is inseparable from the process of mobilization.[23]

Manuel Castells' ideas about the network society have been previously mentioned. In his later work, he presents a very clear example of the connection between virtual space and public space with the Tunisian rebellion. In his book, Networks of Outrage and Hope, he calls this phenomenon "a hybrid public space of freedom": "The connection between free communication on Facebook, YouTube, and Twitter and the occupation of urban space became a major feature of the Tunisian rebellion, foreshadowing the movements to come in other countries".[24]

Recent work has sought to provide alternative frameworks to better address the complexities of communication within social movements. Scholarship using the metaphor of media ecology[25] has focused on resisting reductionism, exploring a multiplicity of media forms, studying the unfolding of using media over time, and recognizing the corporate and state interest and involvement in media.

Identification of the multiple audiences for mobile communication[26] provides another way to better understand the multiple affordances of this media. Identification of activist to activist, activist to mainstream media, and activist to authorities' communications helps to clarify our understanding of power relationships and goals.

Some scholarship on social movement and media[27] has sought to resist technological reductionism by connecting studies of protest movements to established models of mainstream media.[28] While recognizing affordances of social media, this identification of adoption, abstention, attack, and alternatives gives central importance to protesters' intention and motivation.

The protests in Charlotte relied on Black Lives Matter for inspiration, identity, and organization. However, it is important to note that Black Lives Matter is a chapter-based organization that is active through different cities in the United States but does not have a chapter in Charlotte. The protests in Charlotte could be considered as a sister movement that took place in solidarity. Thus, organizations in Charlotte may employ tactics that are not standardized or used by the official chapters of the Black Lives Matter movement.

The protests that occurred in Charlotte in 2016 were spontaneous and had many news agencies around the world following the events. This research aimed to study these events as a case for understanding the relative importance of urban space, the local community, and social media. Social media and information technology played a crucial role in bringing people to the public space. The interviewees provided us with a narrative about which places in the city were used and a rationale for why those places were important. Some places such as the area along Old Concord Road served a role in supporting the family of the victim where the shooting happened. Blocking and taking over highways and interstates served as a way to disrupt the physical flow of the city for higher impact. Many other places were identified as being important due to their strategic location, better accessibility, and proximity to economic activities in the city. Protests during the second and third nights moved to these central areas in Uptown Charlotte.

However, these Uptown areas were not treated uniformly. During the second night of protests, activists purposefully moved from areas around Marshall Park in the periphery of Uptown, to Trade and Tryon which is at the center. These actions caused a direct response from police officers, who were not concerned with the protests occurring in the sanctioned areas of Marshall Park at the margins of the center city. This strategy reveals the policies of segregation that motivated the separation of these two urban spaces. Analyzing the geographic and demographic features around these areas highlighted a gap between these two areas that are near each other but have very different features in terms of demographics and accessibility. Privatization of public space was another factor mentioned in the interviews that could be seen as a strategy to curb public protests. The data analysis showed little to no mention of attempted disruption of a football game.

The interviewees all used social media of different types to diffuse information, to contact friends, and organize and plan future events. However, each social media was used differently. Text messaging was used to virtually connect with friends and acquaintances. Encrypted messaging was used to communicate without the fear of being monitored. Social media such as Facebook with features such as Events and chat groups enabled protesters to organize future events. Some social media such as Instagram and Snapchat that are more multimedia focused were used to share images and videos of protests as they happened. Livestreaming played a crucial role in motivating people by offering the most realistic and real-time snapshot of the protests. Twitter, which is the most public but most restricted platform in terms of content, offered a means for people to communicate with a large audience in a streamlined manner, to spread news quickly, to discuss issues, and to suggest movement in the city.

Indeed, the use of social media is distinguished both by selective use of media (public social media versus private messages), as well as the function that such media plays relative to public space. There were two distinct nights of protest, and social media played an important role in motivating participation in each night and setting the initial locations. Based on an analysis of interviews and tweet data, the tactical shifts in the focus of the protest from the neighborhood of the shooting on the first night and from Marshall Park to the Square at Trade and Tryon on the second were, by contrast, largely directed in person or through private messages.

Even though Twitter allows users to geolocate their tweets, the public nature of the platform does not invite many people to enable this feature. Most of the spatial organization using Twitter happens through text, images, and videos. Analysis of tweet texts allowed for extracting geospatial information. Many tweets in the dataset did not include spatial information; however, by studying the frequency of how each specific place in Charlotte was mentioned, the most important places in the protests identified by the interviewees were found to be also the highest mentioned spatial information in people's tweets.

Social network analysis allowed us to see that users interacted mostly with individuals with similar stances towards the movement. Community detection resulted in two major communities with contrasting stances towards Black Lives Matter. These results support previous examples of echo chambers in social media.[29]

Moreover, the top influencers in the Twitter social network include many news agencies and journalism-related users, as well as politicians and professional activists. This illustrates that during the Charlotte protests, different social media had different functions, and the public nature of Twitter allowed for more news on related and on-the-ground information, while other media might have offered better capabilities in organizing future events. A more in-depth data analysis of other social networks such as Facebook, Snapchat, and Instagram could yield a more holistic view of these dynamics.

The two detected communities treated spatial information differently, a finding supported by a body of research that highlights the significance of location and place on social behaviors and specifically on protests.[30] Furthermore, the two communities show different temporal behaviors. The sympathetic community was the first to massively utilize social media on the first night of protests. The first-night protest happened near the shooting, an area that is mostly residential and is within walking distance of the University. As the activists and protesters gathered in Uptown Charlotte, the protests gained national and international news coverage. The Twitter activity of the sympathetic community remained virtually the same as the first night, while the unsympathetic community had a huge spike. Given our observation that the "against" community contains many non-local users, this spike might be due to a national social media activity in criticism of the protests.

Interviews revealed a narrative about which places in the city were used and a rationale for why those places were important. Some places such as Old Concord Road served a symbolic role to support the family of the victim where the shooting happened. However, taking highways and interstates served as a way to disrupt the physical flow of the city for higher impact. Many other places were identified as being important due to their strategic location, better accessibility, and their proximity to many varied economic activities in the city. Protests on the second and third nights moved to these areas in Uptown Charlotte. Activism is a complex phenomenon. Indeed, in the current atmosphere and with the prevalence of information and communication technologies, activism happens across many different layers.

One of our most salient findings is the highly polarized and segregated opinions of social media users. People who are supportive have stronger connections with each other, similarly, individuals who are against these movements are also strongly connected. Significantly, the structural engagement of the users in these two groups is distinct. Users connected through social media, community, *and* public space are different from those who communicate solely on social media. Detailed spatial information is spread strongly through the former group that supports the movement, while criticism of the protests does not engender spreading spatial information. There is limited evidence from a single study, but this bifurcation of opinions may be a structural property of the nature and density of a user's involvement rather than one determined by content.

This information is not spread only in textual format. One point evident from the interviews was the power of livestreaming on Facebook. The amount of information included in video streams allows for motivating people from around the country to show physical and virtual support to the protests. As these technologies progress, more seamless and realistic information will allow for real-time communication of spatial, emotional, and political information during protests.

The converse of the use of social media to enable public events is the effect that such events have on the frequency and character of social media data. There is clear evidence that during these protests social media activity is correlated with public demonstrations; temporal-spatial events provoke bursts of social media activity. Further, tweet activity by the sympathetic community is heavily influenced by participation in public events, while the unsympathetic community is lower in overall frequency and influenced only by media coverage. Social media can serve as an extension of a discourse begun in public space.

The study of the Charlotte protests showed that different urban spaces indeed have different functions for activists and protesters. Urban space can serve a role for the site of events as well as functional or symbolic spaces in the larger city. This can be reflected in people's narratives and social media data. There is evidence that social media can help to amplify the impact of these protests and transcend the localities of these movements. This is evidenced by the immediate support in our Twitter dataset for the Charlotte protests in Chicago and New York and the participation of people from other cities of North Carolina and the United States.

The Keith Lamont Scott protests shared a number of features with other social movements. The use of multiple social media platforms allowed organizers and participants to maintain privacy for strategic decisions while using others for recruitment and participation. In some cases, surveillance by police forces necessitated the use of encoded information or the adoption of unusual media for planning purposes. The identification of new locations for protest is a critical decision that changes over the course of a movement.

One notable feature of the social media associated with this protest is the effect of distance on the formation of communities during the duration of the protest. Distance as a factor for methods of communication (ranging from in-person to telephones, to telegraphs, to email, to cell phones, to text) has been of interest in geography and urban studies, often centering on the effect of location on the use of alternate media. The changes and shifts in the use of these media (or by the prediction by some that location may be irrelevant) have led to a detailed field of research that can often see the shift of use rather than complete overthrow. This is sometimes referred to as glocalization, a clever contraction of global and local, signifying the conditional and mixed use of media depending on kinship, shared interest, and the impact of in-person events. Studies focused explicitly on social media show variation in the effect of localization on the creation of online networks, some focused on shared interest and some on shared location.

Notes

1 An early version of this chapter was published as "Anatomy of a protest: Spatial information, social media, and urban space." *Social Media+ Society, 6*(1).

2 The Keith Lamont Scott Death Investigation" (PDF). Mecklenburg County District Attorney's Office. November 30, 2016. Archived (PDF) from the original on December 1, 2016. Retrieved November 30, 2016.

3 "Protests Break Out After Man Killed In Officer-Involved Shooting In Charlotte". WCCB-TV. Charlotte, North Carolina. September 21, 2016. Archived from the original on September 22, 2016. Retrieved September 22, 2016.

4 Bell, Adam, Mark Price, & Katherine Peralta. (21 September 2016). "Charlotte police protests: Governor declares state of emergency as violence erupts for second night". The Charlotte Observer. Archived from the original on September 23, 2016. Retrieved September 23, 2016.

5 Welch, Susan, Lee Sigelman, Timothy Bledsoe, & Michael Combs. (2001). *Race and Place: Race Relations in an American City*. Cambridge: Cambridge University Press.

6 Gordon, Cynthia, Marnie Purciel-Hill, Nirupa R. Ghai, Leslie Kaufman, Regina Graham, & Gretchen Van Wye. (2011). "Measuring food deserts in New York City's low-income neighborhoods." *Health & Place, 17*(2), 696–700.

7 Brault, Matthew W. (2012). *Americans with Disabilities: 2010*: US Department of Commerce, Economics and Statistics Administration, US Census Bureau Washington, DC.

8 Dobbins, Dionne, Michelle McCready, & Laurie Rackas. (2016). Unequal access: Barriers to early childhood education for boys of color: Robert Wood Johnson Foundation. Retrieved from http://usa.childcareaware.org/wp-content/uploads/2016/10/UnequalAccess_BoysOfColor. pdf

9 Mohebbi, M., Annulla Linders, & Carla Chifos. (2018). Community Immersion, Trust-Building, and Recruitment among Hard to Reach Populations: A Case Study of Muslim Women in Detroit Metro Area.

10 Miller, Ted R., Bruce A. Lawrence, Nancy N. Carlson, Delia Hendrie, Sean Randall, Ian. R. Rockett, & Rebacca S. Spicer. (2017). "Perils of police action: A cautionary tale from US data sets." *Injury Prevention, 23*(1), 27–32.

11 Khonsari, Kaveh Ketabchi, Zahra Amin Nayeri, Ali Fathalian, & Leila Fathalian. (2010). *Social Network Analysis of Iran's Green Movement Opposition Groups using Twitter*. Paper presented at the Advances in Social Networks Analysis and Mining (ASONAM), 2010 International Conference on. Tufekci, Zeynep & Christopher Wilson. (2012). "Social media and the decision to participate in political protest: Observations from Tahrir Square." *Journal of Communication, 62*(2), 363–379.

12 Carney, Nikita. (2016). "All lives matter, but so does race: Black lives matter and the evolving role of social media." *Humanity & Society, 40*(2), 180–199.

13 These events are by no means exhaustive. We collected these events from our interviews and matched them with this article: http://www.charlotteobserver.com/news/local/article103131242.html

14 Manning, Christopher, Mihai Surdeanu, John Bauer, Jenney Rose Finkel, Steven Bethard, & David McClosky. (2014). *The Stanford CoreNLP Natural Language Processing Toolkit*. Paper presented at the Proceedings of 52nd annual meeting of the association for computational linguistics: system demonstrations.

15 Blondel, Vincent D., Jean-Loup Guillaume, Renaud Lambiotte, & Etienne Lefebvre. (2008). "Fast unfolding of communities in large networks." *Journal of Statistical Mechanics: Theory and Experiment, 2008*(10), P10008.

16 Page, Lawrence, Sergey Brin, Rajeev Motwani, & Terry Winograd. (1999). *The PageRank Citation Ranking: Bringing Order to the Web*. Retrieved from http://ilpubs.stanford.edu:8090/422/1/1999-66.pdf.

17 Juris, Jeffrey S. (2005). "The new digital media and activist networking within anti–corporate globalization movements." *The Annals of the American Academy of Political and Social Science, 597*(1), 189–208.

18 Hardt, Michael & Antonio Negri. (2005). *Multitude: War and Democracy in the Age of Empire.* London: Penguin.

19 Gerbaudo, Paolo. (2018). *Tweets and the Streets: Social Media and Contemporary Activism.* London: Pluto Press.

20 Alterman, Jon B. (2011). "The revolution will not be tweeted." *The Washington Quarterly, 34*(4), 103–116.

21 Earl, Jennifer et al. (2013). "This protest will be tweeted: Twitter and protest policing during the Pittsburgh G20." *Information, Communication & Society, 16*(4), 459–478.

22 Hardt Michael & Antonio Negri. (2005). *Multitude: War and Democracy in the Age of Empire.* London: Penguin.

23 Gerbaudo, Paolo. (2014). "The persistence of collectivity in digital protest." *Information, Communication & Society, 17*(2), 264–268.

24 Castells, Manuel. (2015). *Networks of Outrage and Hope: Social Movements in the Internet Age.* John Wiley & Sons.

25 Treré, Emiliano & Alice Mattoni. (2016). "Media ecologies and protest movements: Main perspectives and key lessons." *Information, Communication & Society, 19*(3), 290–306.

26 Neumayer, Christina & Gitte Stald. (2014). "The mobile phone in street protest: Texting, tweeting, tracking, and tracing." *Mobile Media & Communication, 2*(2), 117–133.

27 Kaun, Anne. (2017). "'Our time to act has come': Desynchronization, social media time and protest movements." *Media, Culture & Society, 39*(4) (2017): 469–486.

28 Rucht, Dieter. (2004). "Movement allies, adversaries, and third parties." In *The Blackwell Companion to Social Movements* (pp. 197–216). Malden, MA: Blackwell.

29 Colleoni, Elanor, Alessandro Rozza, & Adam Arvidsson. (2014). "Echo chamber or public sphere? Predicting political orientation and measuring political homophily in Twitter using big data." *Journal of Communication, 64*(2), 317–332.

30 Carter, J. Scott, Shannon K. Carter, & Mamadi Corra. (2016). "The significance of place: The impact of urban and regional residence on gender-role attitudes." *Sociological Focus, 49*(4), 271–285. Endres, Danielle & Samantha Senda-Cook. (2011). "Location matters: The rhetoric of place in protest." *Quarterly Journal of Speech, 97*(3), 257–282. Gül, Murat, John Dee, & Cahide Nur Cünük. (2014). "Istanbul's Taksim Square and Gezi Park: The place of protest and the ideology of place." *Journal of Architecture and Urbanism, 38*(1), 63–72.

4

HONG KONG

Nowhere has political protest been more sustained than in Hong Kong from 2019 through 2020. These protests are also notable for their shifting locations in urban space and their inventive use of multiple media and social media platforms.

The Hong Kong extradition bill introduced in April of 2019 triggered a series of large-scale protests lasting more than a year. It would have allowed for criminal suspects to be extradited to mainland China under certain circumstances. Opponents feared exposing Hong Kong citizens to unfair trials and violent treatment as well as giving China greater influence over Hong Kong that could be used to target activists and journalists. As noted below, these protests are marked by their intermittent timing and shifting locations within the city. Both of these aspects are due to extreme opposition by the police and the need to hide the time and location of the protest.

The first protest was held at the Central Government Complex on March 15, 2019, followed by events on March 31 and April 28, 2019. On June 9, a march on Hong Kong Island drew a crowd estimated between 3000,000 and 1 million. On June 12, protesters successfully surrounded the legislative council and prevented a second reading of the extradition bill. After the bill's suspension was announced on June 15, a protest drew as many as 2 million people. On July 1, a march by the Civil Human Rights Front drew 50,000 to the legislative council, where protesters briefly occupied the complex.

Protesters later moved to other areas of the city: July 7 on Kowloon, July 14 in ShaTin and New Town Plaza, July 21 on Hong Kong Island, and counterprotests at Yuen Long station. On July 27, protesters again appeared at Yuen Long station and ended in a clash with the police, followed by marches the next day on Saai Wen and Causeway Bay.

DOI: 10.4324/9781003026068-5

Protests followed on August 3 in Mong Kok with some marching to the Cross-Tunnel toll plaza. Protests in Kennedy Town on August 4 in Tai Po, on August 10 in Tai Po spread widely. A general strike was called for August 5 that led to city-wide disruption. From August 12–14 there was a sit-in at Hong Kong airport. A rally by the chief was held in Victoria Park on August 18. Beginning with the Kwun Tong protest on August 24, protesters targeted railway operators. Another set of protests began on the streets of Hong Kong Island on August 31.

Carrie Lam announced on September 4 the withdrawal of the extradition bill. On September 8 protestors marched on the US embassy. A mass protest on September 15 and another on the 21 were met by armed resistance.

Mass protests occurred throughout the city during the first half of October. On October 4, the extradition bill was withdrawn. On October 20 a March occurred on Kowloon Mosque, and on October 23 and 30 on Tuen Mun.

Protesters at Sheung Tak Estate on November 3 led to a city-wide strike on November 11. Conflicts continued at the Chinese University of Hong Kong followed by a siege at Hong Kong University. Following an electoral landslide for District Council, protestors returned to the street in large numbers on December 1 and 8. Popup demonstrations continued through the month and into the new year, focused on major shopping centers. With the outbreak of Covid19 in February, the demonstrations dwindled.

With the announcement that the mainland government would impose new restrictions, a mass protest was called on May 24 in Causeway Bay which continued for several days.

The spontaneous Charlotte Black Lives Matter series of protests in reaction to the shooting of Keith Lamont Scott that happened in three major locations across approximately one week, and the massive multi-city and nation organized Women's March protest that occurred on a single day in important public spaces of many different cities around the world. Compared to these, the anti-extradition protests in Hong Kong are unique, in that they are a series of protests happening across different dates and many locations in the metropolitan area of Hong Kong. Illustration 4.1 shows a timeline of the Hong Kong protest locations connected with arrows denoting temporal continuity. The illustration highlights the movement of protesters from a central business and governmental area of the city to locations that span a large geographic region.

Asides from the unique geographic spread of the protests, protesters' creative usage of various social media platforms is also notable. Throughout the protests, Chinese authorities imposed tight controls and restrictions on social media. However, the use of open platforms on social media made it difficult for the government to identify leaders of the protest since it had no single point of control. For example, protest discussions have utilized LIHKG (https://lihkg.com), a Chinese Language Hong Kong forum, similar to Reddit. LIHKG has multiple users and forums and so the structural search methods are not effective but the tool is efficient because users vote for discussion threads which drive the top threads that people see first. LIHKG

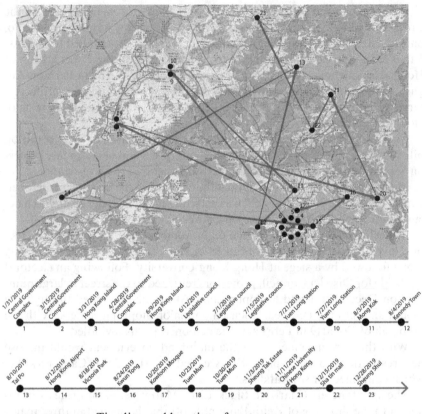

ILLUSTRATION 4.1 Timeline and location of protests.

is a democratic approach to protest promoting horizontal decision-making based on popularity.[1]

Protesters sent messages using group chats and social media platforms with unique affordances. For example, Telegram was used to host message groups of up to 70,000 members and was used for voting next protest locations.[2] Another unique usage of social media was through the dating app Tinder utilizing "several profile pictures with calls for protest and security tips on how to deal with the police".[3] Other apps such as WhatsApp, Uber, Firechat, Pokemon Go, and Line along with the iPhone's Airdrop were used to organize protests. There was extensive posting to chat forums. Activist videos appeared on Douyin, China's version of TikTok. At the same time, in the physical space of Hong Kong subways, commuters received protest invites via Apple airdrop and Bluetooth, both of which are untraceable. Street demonstrations were broadcast via Facebook and the Twitch gaming platform. Pokemon Go was used to organize protests and provide a cover to explain to the police why they were congregating in an urban space. As the skirmishes between the police and protesters escalated and became more violent, severe restrictions were imposed on social media and urban space. Protesters were forced to use an expanding range of media, often in an ad hoc and surreptitious manner. Dating, gaming, and mobility apps allowed them to be a step ahead of the police. Activists switched from text messages (which is easily parsed) to messages using images that are relatively harder to analyze.[4]

Hong Kong protests also had a presence in multiplayer online games. Dean Chan[5] views protest in video games as both a reflection of events in the world and a new field of political and ideological struggle. Hugh Davies[6] discusses a particular version of this strategy used by Hong Kong gamers. They began using virtual private networks in late 2019 to access the Chinese version of *Grand Theft Auto V*, the first title to use an open game world. This allowed gamers to manipulate the actions and appearance of the avatars and to create familiar Honk Kong urban space within the game. Hong Kong players customized their avatars as protesters and tossed petrol bombs, vandalized train stations, and attacked the police. Chinese players responded by creating avatars as riot police and clashed with the protesters.

A post on the *u/anagoge* subreddit included a photo of the police in riot gear at a protest with the caption, "This photo of the Hong Kong protests looks straight out of a video game". A series of posts quickly followed comparing this image to numerous existing games. Just three weeks later demo videos were posted of *Liberate Hong Kong* (Illustration 4.2). The work of an unknown team of developers, it strongly resembles the image

Subreddit photo of Protests

Liberate Hong Kong Video game

ILLUSTRATION 4.2 Photo of riot police and Hong Kong video game.

posted originally and included urban spaces that had been sites of conflict. Within the game, players were assigned tasks including collecting tear gas containers, avoiding arrest and rubber bullets. The possibility of *Liberate Hong Kong* connected with ongoing discussions of Massive Multiplayer Online Role Play Games (MMORPG) in Hong Kong and their ability to brainstorm new and creative tactics such as using umbrellas to neutralize police weapons or spark demonstrations in 40 cities simultaneously.[7] The use of gaming to create an overlapping of virtual and physical space as part of a protest is unusual. Gaming can provide a testing ground, an organizing forum, and a statement of principles for the protest. Neither the world of the game nor the physical world is replaced or obscured. The game can be a dress rehearsal for the real protest, but just as certainly it is also another venue to struggle for the upper ideological hand.

Social media was critical in communicating with the local community. Besides innovative usage of various forms of social media to communicate and organize with other protesters, the long-term organized nature of the Hong Kong protests required a much more organized way of communicating information with citizens. The r/HongKong subreddit includes a "mega thread" on anti-expedition protests. This English thread with more than 19,000 upvotes includes a list of information about Hong Kong and the protests. Most of the information from this thread is translated from other social media platforms written in Cantonese. Within the thread, there are several links to other resources such as links to other Reddit posts answering questions such as "should I go to Hong Kong or Not?", apps such as CityMapper,[8] a Reddit live feed that constantly propagates information about protests.[9] One notable link includes a Telegram channel where "live information" is posted about the protest. As a unique case of long-term continuous protests, it is important to understand how real-time information is communicated with users, what kinds of information are shared with them, and to what extent this information is spatial and overlaps with urban spaces.

Social Media Analysis

To explore these questions, more than 26,000 messages from the public Telegram channel StandWithHongKong (https://t.me/StandwithHKlive) were collected from dates X and Y. The StandWithHongKong channel describes itself as follows: "We provide English #HongKong protests updates (and now also #coronavirus updates), Message translated from existing Chinese channels. Live updates mostly only during weekends. This channel provides info only. We don't promote/organise any activities".

In order to uncover this unique case of live feeds about long-standing protests, text data from messages was gathered and analyzed. First, using NER, names of places were extracted from the text of messages using the Spacy python library. The top places entities were geocoded using the open street maps geocoding API (Nominatim). This allowed for approximating the locations where Telegram messages are referring to. Through this process, a total of 1194 unique entities were extracted in the Hong Kong region. The highest mentioned locations were MongKok with more than 1500 and Kowloon with more than 500 mentions. Illustration 4.3 highlights a map of Hong Kong overlaid by a grid showing the number of times places within each grid cell have been mentioned by the StandWithHongKong Channel. The sheer spread of different places that are mentioned by the channel shows a direct contrast to more spontaneous protests within other places such as the Black Lives Matter protest in Charlotte, North Carolina where mentioned places by protesters are limited to only a few street intersections and famous places.

The next step in analyzing the content of messages is to understand the general categories of information that are produced by this channel and whether and how each category is associated with various regions of Hong Kong. By categorizing the messages, a general understanding of the relationship between social media and local communities in long-term protests and unrest can be achieved. More specifically, two specific questions are asked: (1) what are the different kinds of information that is transmitted to citizens/protesters? (2) Do different places elicit different kinds of information for audiences?

To find different categories of information, Top2Vec[10] topic modeling was used. With minimal supervision, Top2Vec clusters documents into different

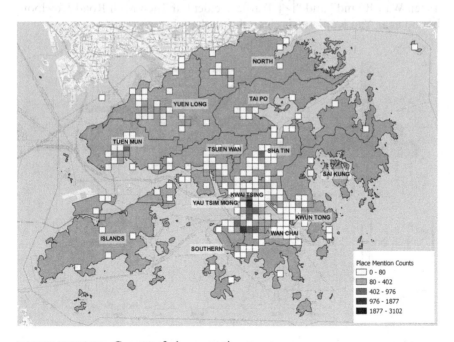

ILLUSTRATION 4.3 Counts of place mention.

groups using pre-trained embeddings from BERT (other embeddings are provided with this method). Initially, the model produced 184 topics for the 26,000 documents within the dataset. Using hierarchical merging of smaller topics to similar larger topics, 15 general topics were finally extracted. The top 20 messages most closely related to each topic were examined by the authors to determine the nature and characteristics of each topic and whether some topics are related to similar concepts/categories of information. Categories of messages were extracted by the authors by identifying themes from the topics extracted from the 26,000 messages. Special focus was given to messages with the top ten mentioned places which amounted to a total of 5000 messages. (See Illustration 4.4).

With words such as traffic, citybus, highway, bus, tunnel, the most prominent topic within the corpus were messages that relayed specific traffic information of interest to audiences. Such messages included information about road closures, traffic congestion, cause of disruption, and occasional messages to avoid; for example, *"Due to road conditions, 59 routes of New World First Bus and Citybus are suspended, 45 routes are having traffic diversion…"*. The traffic-related messages often include information about several streets and roads, and use different emojis to communicate blockage, or safety, and guide users to avoid certain areas of the city.

Some topics with words including roadblock, highway, road, and roads grouped messages that generally included information about closures. Messages often include reasons for road closure and whether a specific direction is being affected. For example, "2043 Road block at Shing Mun Tunnel Tsuen Wan Bound" and "☑ Traffic accident at Tuen Mun Road Kowloon bound near Yau Kam Tau is cleared #roadcondition".

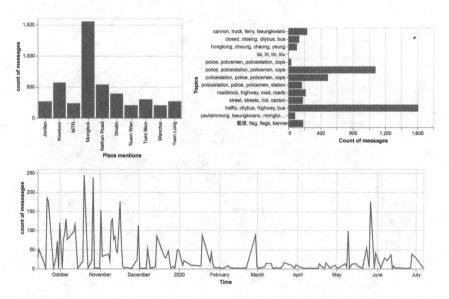

ILLUSTRATION 4.4 Top topics, temporal activity, and top place mentions.

Many of the traffic-related information communicate specific information about the public transportation systems and bus routes that are suspended or ones that have diversion such as *"Route(s) suspended 1, 2X, 5B, 5X, 8, 8P, 8X, 10, 11, 12, 12A..."* or *"Tram services between Shau Kei Wan Terminus and Finnie Street are suspended"*. Other traffic-related messages start with phrases such as "Stations to avoid" or "Stations you should stay away from", which specifically instruct protesters and citizens to avoid specific areas, streets, or public transit routes.

Social media proved to be very adept at detecting police presence and movement. Several extracted topics included information about the presence and the behavior of police in various parts of the city. Many of these messages included descriptions of the types of police, as well as a specific address where they were observed and their general direction of movement. For example, *"#MongKok 2338 3 EU 6 carriers 2 private cars moving towards Mong Kok along Nathan Rd"* and *"6 police cars toward Kwai Hing from Kwai Fong police station"*.

Mentions of riot police and water cannons were also among the prominent elements within the police-related messages from the live channel. For example, "2212 #Mongkok #NathanRoad Water cannon vehicle shot water, heading to #YauMaTei", "2311 Water cannon no. 1, 2 armored vehicles, 10+ police cars from Nathan Rd towards TST", and "0131 #TinShuiWai 9 riot police walking from Tin Shui Wai West to Tin Yiu LTR Station".

Messages also sometimes included other information about the series of events and actions of the police. These messages included keywords such as "stand by", "charging towards", "entering", "on the ground", and "disperse and charge". For example, *"#TinShuiWai: "Sing with you" activities in Ginza Plaza undergoing. Be alert of police to enter into shopping mall to disperse and charge citizens"* and *"#TaiPo, 1910 riot police entering Tai Po Mega Mall, Police at #TaiPo Mega Mall as there are shops being vandalised"*.

Colored flags emerged as a topic with words such as flag, flags, and banner emerged from the live update channel. Examples of such messages include: "2201 40–50 riot police charge along Nathan Rd towards Yau Ma Tei direction with blue flag", "1906 Orange Flag at Mongkok Police Station", and "black flag again at #NathanRd #ArgyleSt". These colored flags are symbols used by Hong Kong police to communicate various messages to protesters. For example, a red flag is used to communicate "Stop charging or we use force", while a black flag communicates "Warning – tear gas", and an orange one is used to communicate "Disperse or we fire". The messages that alert users about observations of different kinds of flags are accompanied by stickers of a dog that resembles the doge meme holding the corresponding flag. Occasionally, the messages also include a picture of the police force holding the mentioned flag. Illustration 4.5 shows the flag stickers used by the StandWithHongKong live channel. These flags are likely used for distinguishing these special messages about flags from the rest of the more usual messages about traffic or police presence.

ILLUSTRATION 4.5 Colored police flags in Telegram to mark clashes of police and protesters.

Case Study: November 11, 2019 Protests at Chinese University of Hong Kong

The protests centered at the Chinese University of Hong Kong on November 11, 2019, were covered extensively by the western media.[11] Over several days, protesters held protests within the CHUK campus while the police used tear gas and water cannons to disperse or contain the protests (see Illustration 4.5). Focusing on the messages from the StandWithHongKong channel published during the protests allows for a more detailed spatial view of what information is propagated to protesters and citizens.

Police presence was covered extensively by mainstream media. The majority of the descriptions were related to the actions of the police in the vicinity of the university campus. For example, excerpts from the Associated Press' coverage[12] such as "surrounded the area Sunday night and began moving in after issuing an ultimatum for people to leave the area", and "Riot officers broke into one university entrance before dawn Monday as fires raged inside and outside, but they didn't appear to get very far. Fiery explosions could be seen as protesters responded with gasoline bombs", describe the police forces' activities in fine-grained descriptive detail.

Live messages about the same protest are shorter and more to the point and often describe a single event in a specific location. For example, "*#CUHK 2322 water cannon and armored vehicle moving towards Science Park direction along Tolo Highway*", "*#CUHK, 2307 water cannon and few dozens police at Fo Yin Rd roundabout, pls be careful if leaving via Tai Po*", and "*#WongTaiSin 1123 Bridge outside Hsin Kuang Centre -10 blue uni police standby*".

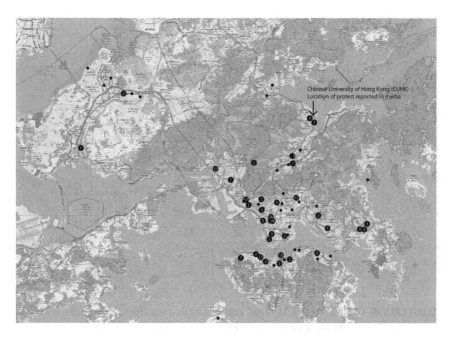

ILLUSTRATION 4.6 Locations of messages noting Hong Kong police during the CHUK protests.

Furthermore, messages sent by the live channel that relate to police presence do not solely focus on police presence around the university campus but all over the metropolitan area (See Illustration 4.6). For example, on the same day as the CHUK protests, the channel covered police locations and behaviors for places such as Tsuen Wan that is located at the opposite end of the city "*#police 1509 #TsuenWan 1 carrier and 2 EUs moving around at Tai Ho Rd, 7 cops with guns on ground*", and Central Park that is on the south side of the city, "#police many police vehicles going towards Central Park #Shatin from Yi Ching Lane Royal Park hotel".

Unlike mainstream media that rarely go into specific details about the protests such as flag colors shown by the police force, the channels' coverage of colored flags by police is also extensive and covers a large span of the geographic area. During the protest days, mentions of orange flags (disperse or we fire) were used in a message covering CHUK protests. Mentions of orange flags in other parts of the city, as well as several blue (disperse or we use force) and black (warning tear gas smoke) flags were made for other parts of the city (See Illustration 4.7).

Finally, there were a large number of messages during the CHUK protest day that covered traffic, road closure, and public transportation information. Messages were longer and contained detailed information about routes that were closed or ones that have traffic. Illustration 4.8 shows the large number and spread of messages about this topic around the Hong Kong

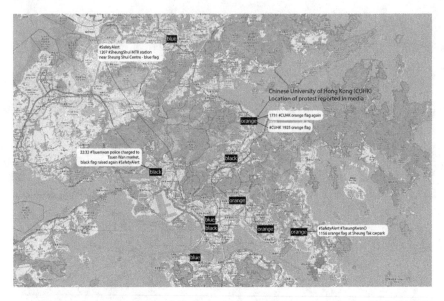

ILLUSTRATION 4.7 Mentions of flags and example messages during the CHUK protest day.

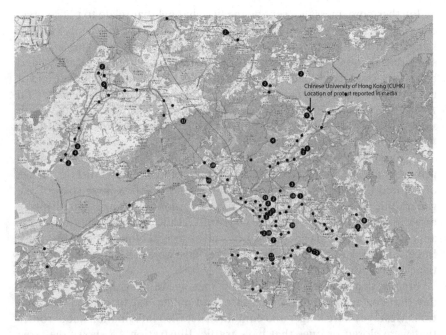

ILLUSTRATION 4.8 Messages containing traffic, road closure, and public transportation during the CHUK protests.

metropolitan area. The following is an example of one of the messages covering Tsuen Wan area:

"⚠20:12 #TsuenWan #MaOnShan #TrafficInfo #KMB

`the following suspended✗:`
30, 31, 34, 43, 43B, 238M, 243M, 235

The following routes will be diverted (Omit Tai Ho Rd)⚠
30X, 40, 41M, 42M, 43X, 49X, 51, 53, 73X, 234A, 234B, 278X, 290, 290A, 290X, E31, E32A

43X

will turnaround at Sai Lau Kok Rd
`Omit the following stations:`
`✗Chung On St, Tai Ho Rd, Tsuen Wan West`

49X (both bounds)
`✗ omit all the stops along Tsuen Wan"`

Global Response

After seeing the specificity of the spatial protest-related information propagated in the live channel communicated to protesters, it is important to contrast this coverage by the Tweets posted about Hong Kong protests (including #hongkongprotest hashtags). This brief qualitative analysis will primarily allow for differentiating the more global social media activity on Twitter from the local live feeds of Hong Kong protesters. Using Top2Vec, the same Topic Modelling procedure used to summarize the live feed messages, more than 1,00,000 tweets were collected from April to October 2019 and categorized in 15 overarching topics. By reading the top related tweets on each topic, several themes were extracted.

Social media was critical in raising awareness about the protests and their demands. Several topics with words including hongkong, hk, china, and mainlanders related to tweets that aimed to raise awareness about the protests for the global community. Tweets such as "please stand with Hong Kong people" and "Please let Hong Kong be Hong Kong!", and "Oh my god! Hong Kong really need help 😢" asked for protesters' demands to be heard. Other tweets included hashtags such as #hongkong #HongKongProtests #PoliceBrutality #antiELAB #NoToChinaExtradition to trend the protest-related tweets. Some topics discussed China's role in the series of events. Some tweets such as *"China Is Waging a Disinformation War Against Hong Kong Protesters"* and *"How China is framing the Hong Kong protests"* discussed China's usage of news and media to paint a specific picture about the protesters.

Another important function of social media involved emotional tweets and strong tweets with mentions of democracy. One topic included strong emotional words such as *shameful, omg, heartbreaking,* and *thanks.*

The tweets within these topics described hopes for change, thankfulness towards supporters. For example: *"Thank you for your support to Hong Kong. We will never stop to fight for freedom..."*, *"This is a shame for Hong Kong"*, *"Thank you for supporting #HongKongProtests. We will keep fighting..."*.

There were also a series of tweets discussing concepts such as democracy and freedom. For example: *"Please stand with Hong Kong. We are fighting for freedom and democracy. Ty so much"*, and *"People will never stop fighting for FREEDOM and DEMOCRACY in Hong Kong"*. Some tweets used hashtags such as #chinazi that connected china with the Nazi regime: *"Thank you ☖ Stand strong against #ChiNazi! #FreeHongKong #HongKongProtests"* and *"please stand with us #HongKongProtests #HongKongers no chinazi"*.

Social media was a method of discussing protest events and police behavior. Several topics with words such as protesters, manifestants, and demonstrators described general protest activities in Hong Kong. Many of the tweets were like the news and discussed the events in general terms. For example: *"Massive peaceful protests in Hong Kong against the #ExtraditionBill"* and *"Thousands of anti-government protesters march through a popular shopping district in Hong Kong"*. While some tweets were more detailed and mentioned specific neighborhoods and places in Hong Kong and provide more details about the protests and their clash with police. For example, *"Protester regrouped on queensway and police hv been firing tear gas nonstop"*, and *"Anti-govt protesters fight with Beijing supporters in Hong Kong mall"*.

Police brutality and behavior was also another major theme within the tweet dataset. Several topics with words such as policemen, police, and teargas described various ways police clashed with protesters. For example: *"Police throw tear gas into a crowd of protesters in #Hong Kong"*, *"Tear gas fired at TKO, where protesters are throwing bricks at the police station. Protesters are chanting"*, *"Another round of tear gas at #KowloonBay as protesters combat with police again. #HongKongProtests"*, and *"Chaos on Hong Kong's MTR network as police chase protesters into stations and beat people"*.

Protests and Spatial Information

Social media, the local community, and a global audience were all important for the Hong Kong protests. Anti-extradition protests in Hong Kong received massive local and global coverage from social media, as well as mainstream news media around the world. Topic modelling and qualitative analysis of Twitter data showed that the global coverage of the protests had a primary focus on pulling the global audiences' attention to the Hong Kong protests. Many tweets were emotional and discussed the protest events within the context of protesters' fights for democracy, as well as their fears of dictatorship. While others specifically tried to communicate the state of events, sometimes with specific mentions of where the protests occurred. Police brutality and their usage of tear gas and other kinds of excessive force was a major theme within the tweets.

On the other hand, live message feeds on Telegram had a clear focus to communicate information about the events in the city to the local community. The information was often specific. Messages regarding police described the time, location, and general direction of the observed police. Moreover, there were mentions of police gear such as water cannons and personnel carriers. These mentions of police often were accompanied by hashtags such as #safetyalert or #avoid.

Even though there might be an overlap in the information shared on Twitter and Telegram such as mentions of major protest events and descriptions of the specific place they occurred in, live feeds of the StandWithHongKong Telegram channel highlights a much tighter link between the local community and urban space. Not only anti-protest police observations were reported all around the city but mentions of road closures, heavy traffic, and alterations in public transportation schedules are also signs of how information about the functional aspects of the city is important and essential for the protesters. Moreover, mentions of police flags which is information likely only important to individuals who are navigating the streets of Hong Kong during the protests is an indicator of how social media plays an essential role in the activities of protesters.

Juxtaposing Telegram live messages with tweets is another important point that has recurred in the studies of various aspects of social media. Twitter might be space to describe emotions, call for actions, and gather a large audience to hear the demands and struggles of activities, while Telegram messages are broad and used as a tool for making cities hyper-legible for protesters who are trying to make demands and also stay safe. Together, social media platforms and other technologies with social features create a new hybrid space that links local communities with urban spaces they aim to occupy and make their own.

Notes

1 Purbrick, Martin. (2019). "A report of the 2019 Hong Kong protests." *Asian Affairs, 50*(4), 465–487.
2 https://www.bbc.com/news/technology-48802125.
3 https://www.businessinsider.com/hong-kong-youth-using-tinder-and-pokemon-go-organize-protests-2019-8.
4 Zhao, Jingyi. (2017). "Hong Kong protests: A quantitative and bottom-up account of resistance against Chinese social media (sina weibo) censorship." *MedieKultur: Journal of media and communication research, 33*(62), 28-p.
5 Chan, Dean. (2007). "Dead-In-Iraq and the spatial politics of digital game art activism." In J. Tebbutt (Eds.), *Australian New Zealand Communication Association Annual Conference*, Communication, Civics, Industry, Australia and New Zealand Communication Association and La Trobe University, Melbourne, Victoria; Chan, Dean. (2009). "Beyond the 'Great Firewall': The case of in-game protests in China." In Larissa Hjorth & Dean Chan (Eds.), *Gaming Cultures and Place in Asia-Pacific* (pp. 141–157). New York: Routledge.
6 Davies, Hugh. (2020). "Spatial politics at play: Hong Kong protests and videogame activism." *Proceedings of the 2020 DiGRA Australia Conference,*

February. http://digraa.org/wp-content/uploads/2020/02/DiGRAA_2020_paper_46.pdf

7 Chiu, M.M. (2019). "Are video games making Hong Kong youths delinquents, loners... or better protesters?" *Hong Kong Free Press*. https://www.hongkongfp.com/2019/09/15/video-games-making-hong-kongyouths-delinquents-loners-better-protesters/

8 https://citymapper.com/chicago?lang=en.

9 https://www.reddit.com/live/133sixros7tu5.

10 https://github.com/ddangelov/Top2Vec.

11 https://www.nytimes.com/2019/11/18/world/asia/hong-kong-protests-university.html and https://apnews.com/article/1575c7c2ce5c4cc3a572741b1e806320.

12 https://apnews.com/article/1575c7c2ce5c4cc3a572741b1e806320.

5

WOMEN'S MARCH

On the election night of 2016, upset by the outcome of the US Presidential election, Teresa Shook created an event page on Facebook for a hypothetical protest March. The next morning, she was shocked to have received 10,000 replies. Experienced organizers offered to help and two days later a steering committee was formed.

Facebook was the center of advertising and planning for what became the Women's March on Washington.[1] Facebook's ability to scale almost immediately to organize and share a very large, public event made it possible in only 60 days. What began as shared discontent with their president-elect and fear for their rights to autonomy, votership, and equality grew to become an international movement for gender equality. The focus of the planning was for a very open and inclusive event with a total of 100 cosponsoring organizations. The organizers wanted a prominent and central public space and had thought of the Lincoln Memorial as their first choice for the March, but a scheduling conflict led to a different site at the National Mall. As a consequence of the early organizing, Sister Women's Marches were held in 300 American cities and approximately 300 more cities around the world at town squares and city centers. The Women's March coordinated hundreds of charter buses to provide transportation to and from DC for participation. The Washington Metropolitan Area Transit Authority reported record numbers of passengers. Despite the short window, organizers were able to provide sanitation services, volunteers to run the event, and maps and other information for the protesters.

The Women's March was held on the national mall on January 22, 2017, the day after the inauguration of President Trump. The March in Washington DC drew three times the original estimate of 200,000 marchers, while "Sister marches" occurred in more than 600 locations worldwide involving 4 million additional Marchers.[2]

DOI: 10.4324/9781003026068-6

There was extensive coverage of the March by the broadcast media afterward. But it was social media that played a central role before and after the event in publicizing the Women's March, highlighted using the hashtags: #WomensMarch and #WhyIMarch. Social media also enabled an individualized focus that figured importantly in subsequent media reports. Iconic "pussy hats" (referring to Trump's infamous on-camera asides) were prominent features of the March.[3] Participants posted photos of hand-drawn posters as a proxy for tweets of individual political opinions, providing a series of personal headlines.

For the Washington Women's March, the importance of the location was a given from the beginning. The Mall had been the site of historic marches and protests including the Bonus Army of veterans in 1932, the Martin Luther King March in 1963, protest marches against the Vietnam War between 1966 and 1973, and the Million Man March in 1993. The importance of the mall as a national symbol and its timing on the day after Donald Trump's inauguration coincided to validate its importance as an event. The other women's March in other cities across the country and the world likewise chose important civic urban spaces. The importance of the symbolic urban space has been discussed by Castells and has elements in common with the Honk Kong and Black Lives Matter case studies. Additionally, the posting of social media over time (particularly on the day of the event) shares aspects with many other events.

Twitter data was collected from across the nation during the period of the Women's March. Approximately 2.5 million geolocated tweets were collected and sorted selecting for references to hashtags connected to the March. Using sentiment analysis, locations were mapped and scored for positive or negative reactions. Some variation was found, but the overwhelming sentiment toward the March was positive.[4]

In another sense, the organization of the March was a remarkable achievement that used social media almost exclusively. There was not a single mention of the Women's March on Network television until two days before the event.[5] All the publicity and organization took place on social media. The Facebook community grew from a single post on the night of the election to 15,000 members on the eve of the March. This event was a clear demonstration of the ability of social media to replace mass media under some circumstances.

Location Matters

A salient feature of the Women's Marches is the posting of images and written comments on social media. Kitcher[6] points out the emergence of Instagram as the primary social media at the time of this event. She also points out the use of colors (mostly pink) and visual symbols as prominent aspects of the users' expression. During the March, the @womensmarch Instagram account was tagged 52,000 times, and the #womensmarch and #womens-march2017 hashtags 600,000 times.

Einwohner and Rochford[7] extend this discussion of Instagram to the period after the Women's March. Noting that social media analysis has usually been focused on its role in organizing protests, they argue that the role of social media for individual participants is best understood using performance theory. This holds that the participation in the March was essentially a political performance and, therefore, a "showing doing", a "pointing to", or "underlining of" an important activity. The data for their analysis is 134 Instagram posts randomly selected from event organizers and individual participants. Using a form of grounded theory, a coding scheme adapted from performance theory enabled an analysis of the differences.

An analysis of the protest signs created for the March[8] began with an understanding that social media makes collective action possible. This form of collective action is not reliant on formal advocacy organizations that control action top down, but rather on a diffuse organization operating at a personal level and "crowd sourcing" through a loosely tied interpersonal network. The signs are extracted from the photographic context and treated as documents that can be understood using thematic analysis, a form of grounded theory. These themes are identified as unity, women resistance, appropriating vaginal language, criticizing Trump, and redefining feminism.

Instagram has become central for many users of social media.[9] Several aspects are critical to our understanding of how it shapes visual culture. Photographic self-representation is central to Instagram and allows for an expanded range of the photographable often associated with young women. These photographs are also typically set in a specific sociocultural milieu. These factors combine to create the potential for a powerful tool that has the potential to comment on social, cultural, and political.

To gain a wider understanding of how performance theory might clarify the relationship between urban space and social media, both a larger body of data and a new method of analysis were required.

This chapter uses the Women's March Twitter dataset[10] that collected a total of 7,275,228 tweets between December 19, 2016, and January 23, 2017. Images were downloaded from a subset of geolocated tweets, image embeddings were extracted from the Imagenet[11] model and K-Means cluster methods were used to cluster photographs. A total of four groups resulted from the analysis (see Illustration 5.1).

The meaning of these groupings required interpretation by the authors. Cluster one that is labeled "portraits" and focuses on individuals or a small group of protesters. These photographs were taken from a frontal view, the subjects engaged the camera directly and sometimes used signs or costumes as props. They provide a personal record of participation.

ILLUSTRATION 5.1 Instagram photo clusters #WomensMarch.

Cluster two, labeled "message" focuses on the individually created signs. For this cluster, the signs were at the center of the photographs, although always with the context of the people who are proudly carrying them and the setting of the March. These photographs of single signs are proudly displayed, photographs with multiple signs merging into the crowd, as well as photographs of a display of posters left as a memorial of the March. Previously published posters from the March are presented in a manner that isolates them from the context of the March and presents them as clever graphic and editorial statements.[12] This presentation emphasizes aspects of the posters that are more abstract and isolated than their representation in the photographs.

Cluster three labeled "crowds" focuses on the actual March, the assembly of the huge crowd moving in unison. The posters are still present, but their legibility is reduced to floating ciphers floated on the sea of protesters. These are the moments when the crowd takes on a collective life, protesters moving both individually and as one single being.

Cluster four labeled "urban setting" contains photographs that have a prominent role for buildings that define a city image or a public setting. Examples in these photographs included the Capitol Building and the White House in Washington. In these photographs the individuals and the crowd are less distinct except as a part of urban space.

The Women's March provoked a huge outpouring of activity during the protest in 2017 in Washington of another of the 400 sites worldwide. Instagram posts had continuous activity long after the event in 2017. A timeline of 550 photographs posted on Instagram from 2017 through 2020 is shown in Illustration 5.2. There is activity using this hashtag during the entire four-year period. This posting is heaviest during the first year after the March in 2017 showing sustained activity with peaks in the three months after the event, again in November on election day, and on the first anniversary of

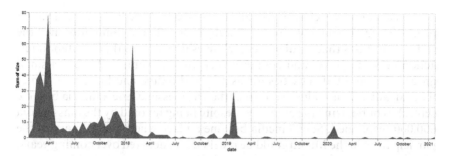

ILLUSTRATION 5.2 Timeline of Instagram posts to #WomensMarchWashington 2017–2021.

ILLUSTRATION 5.3 "Performance" Instagram posts to #WomensMarchWashington 2017–2021.

the March. Posting continued for the next three years at a reduced level, especially on the anniversary of the March and near election day each year. Events on social media are more typically marked by a sudden rise and a short "tail" of activity, but Women's March shows recurrent activity. This finding is consistent with the idea of performance mentioned earlier.

Over those four years, the photographs posted on Instagram develop a wider range of focus. Content arose that was tailored more specifically to the posting format. Political organizing was presented in messages often featuring slogans or calls to action similar to mass media advertising. Others developed sales opportunities for hats and t-shirts that would appeal to March veterans.

In our sample data set, 40% of the postings (either new posting or re-posting) continued to be photographs showing participation on the day of the March discussed earlier through cluster analysis. Illustration 5.3 shows these posts arranged in chronological order. These illustrate the importance of having been with the crowd of protesters on that urban stage. Although there are many methods that politics is carried out such as writing, discussion, or lobbying, being physically present has a powerful impact on memory and engagement.

Like other protests studied in this book, the Women's March conforms to a remarkable degree to the insights of Manuel Castells.[13] In *Networks of Outrage and Hope*, he claims the importance of social media, urban space, and a dedicated local community to the success of protest movements and new forms of social change. He argues that social media is critical to the process of citizen empowerment, whereby in society the configuration of the state institutions that regulate people's lives depends on this constant interaction between power and counterpower. In the Women's March, a small group of committed activities forms a hybrid online and local community that can advance its values and strengthen its message. Social media became central to publicizing the event and despite being ignored by the mass media 600,000 responded. Social media became a place where the memory of the participation in the March could be recalled and debated over coming years.

Notes

1 Lapowsky, Issie. (2017). "The women's March defines protest in the Facebook age." *Wired Magazine.* Accessed from: https://www.wired.com/2017/01/womens-march-defines-protest-facebook-age/
2 Goodman, Amy. (Host & Exec. Prod.). (23 January 2017). "Women's March on Washington: Historic protest three times larger than Trump's Inaugural crowd." *Democracy Now!* Retrieved from https://www.democracynow.org/2017/1/23/womens_march_millions_take_to_streets
3 Walker, Rob. (2017). "The DIY revolutionaries of the Pussyhat Project." *The New Yorker.* Accessed from: https://www.newyorker.com/culture/culture-desk/the-d-i-y-revolutionaries-of-the-pussyhat-project
4 Jessica Gantt-Shafer, Cara Wallis, & Caitlin Miles. (2019). "Intersectionality, (Dis)unity, and processes of becoming at the 2017 women's March." *Women's Studies in Communication, 42*(2), 221–240,
5 https://www.washingtonpost.com/lifestyle/style/how-mass-media-missed-the-march-that-social-media-turned-into-a-phenomenon/2017/01/21/2db4742c-e005-11e6-918c-99ede3c8cafa_story.html.
6 Kitch, Carolyn. (2018). "A living archive of modern protest." *Memory-making in the Women's March, Popular Communication, 16*(2), 119–127.
7 Einwohner, Rachel L. & Elle Rochford. (2019). "After the March: Using Instagram to perform and sustain the women's March." *Sociological Forum, 34*, 1090–1111.
8 Weber, Kirsten M., Tisha Dejmanee, & Flemming Rhode. (2018). "The 2017 women's march on Washington: An analysis of protest-sign messages." *International Journal of Communication, 12*(2018), 2289–2313.
9 Caldeira, Sofia P., Sander De Ridder, & Sofie Van Bauwel. (2020). "Between the mundane and the political: Women's self-representations on Instagram." *Social Media+ Society, 6*(3), 2056305120940802.
10 Littman, Justin & Soomin Park. (2017). "Women's March Tweet Ids", https://doi.org/10.7910/DVN/5ZVMOR, Harvard Dataverse, V1.
11 Krizhevsky, Alex, Ilya Sutskever, & Geoffrey E. Hinton. (2012). "Imagenet classification with deep convolutional neural networks." *Advances in Neural Information Processing Systems, 25*, 1097–1105.
12 Weiner, Samantha & Emma Jacobs, eds. (2017). *Why I March: Images from the Women's March Around the World.* Abrams Image.
13 Castells, Manuel. (2015). *Networks of Outrage and Hope: Social Movements in the Internet Age.* Hoboken, NJ: John Wiley & Sons.

PART 2

Commerce

6

MOBILE FOOD TRUCKS[1]

Mobile food truck vendors that emerged in 2008 are defined by their high-quality and often high-price food served in stylishly branded and fully equipped catering trucks to patrons desiring a new type of outdoor food experience. The rapid growth of these food vendors can be linked to a variety of converging factors. The economic downturn of 2008 and the subsequent decrease in consumer spending and consumer confidence shifted demand away from the broader food-services sector and toward affordable food options.[2] Aspiring chefs and culinary students, who enjoy experimenting with culturally diverse cuisines and who are equipped with the knowledge of the restaurant industry, seek success similar to the popular Los Angeles-based Kogi BBQ food trucks, known for serving Korean–Mexican tacos. Young culinary entrepreneurs and established chefs, who may have difficulty maintaining their restaurants, find starting a mobile food business more financially feasible and flexible. Last, the lack of prior mobile vending precedents and loose municipal ordinances initially allowed vendors to easily navigate urban areas.

Mobile food trucks in Charlotte, NC – the focus of this chapter – have an established mobile food truck scene that began with a grassy parking lot used for farmer's markets in the up-and-coming Historic South End district. In 2010, the first weeks were experimental and slow, hosting just four trucks but grew rapidly in the spring months of 2012. "Food Truck Friday" became the food truck event in the city hosting 20 trucks, each serving between 300 and 400 tickets in four hours.[3] In the succeeding years food trucks in Charlotte have populated Charlotte's abundant office parks, university campuses, and nightlife areas downtown. Although food trucks in Charlotte emerged a few years after they were common on the west coast, over time the industry has evolved with the help of new commissaries.

DOI: 10.4324/9781003026068-8

Information technology is critical for these mobile food trucks, allowing vendors to communicate without sole reliance on word-of-mouth communication, physical proximity to customers, or repeated presence. Information technology allows vendors to communicate and exchange real-time information with patrons as well as to mobilize their business activating urban areas with a predictable crowd. Unlike prior forms of vending that relied on word-of-mouth communication, music from the vehicle, physical proximity to dense foot traffic, social media, etc. affords vendors the ability to instantly notify large numbers of people well in advance and respond to their requests. This dynamic allows vendors to create a demand online consisting of a select population that is well versed in contemporary food culture.

The speed and usability of information technology to promote, connect, and expand vendor operations across cities amplified the growth of the vending industry. Vendors, customers, and advocacy groups use social media platforms (e.g., Twitter, Facebook), smartphone applications that offer real-time tracking of trucks (e.g., TruxMap, Food Truck Fiesta, Foursquare, Road Stoves GPS, and Truck Spotting), smartphone payment applications (e.g., Intuit's GoPayment and Square), photography and video platforms (e.g., Instagram, Vine), as well as blogs and business and food review websites (e.g., MobiMunch, Yelp, FoodTrucksIn, and Urbanspoon). These tools, which together create a media ecology, compliment the nimble and flexible business models of vendors by providing an instant way to build demand as opportunities emerge.

Twitter initially became the most popular social media platform among vendors and still predominates today. Structured on a microblogging framework that allows sending of 140-character messages or 'tweets', vendors use Twitter as a free mass-marketing tool to communicate to a localized audience their latest or future locations, daily menu items, or if they are out of service. Vendors also use Twitter to choose locations on their daily route and to understand the locations of their fellow vendors. Customers find this real-time information helpful in locating their favorite trucks and menu items. In an industry reliant on mobility, Twitter provides a virtual infrastructure to communicate real-time information and assures vendors a sufficient customer base in a variety of public and private underutilized locations.

Structured upon both physical mobility and continuous online communication, mobile food vending today poses interesting questions about the interrelationship between urban space and social media. How does social media inform social practices in urban space and what new spatial and temporal relationships develop as a result? Tweets, posted on the social media website Twitter, provide a lens to better understand the multiple roles and functions of real-time information among vendors as it informs urban settings.

Regarding urban space, the unexpected emergence of food trucks is influenced by their position as less powerful actors in the design and planning of the city. Considered to be an unplanned and locally driven, bottom-up urban activity, food trucks often conflict with established regulations that control the type and frequency of activities in urban space. To understand mobile food trucks, it is imperative to understand how space is socially and politically constructed.

Marxist philosopher, Henri Lefebvre is well known for his views on capitalist urban development and transformation throughout the 20th century. His conceptual frameworks of the *right to the city* and *oeuvre* provide a way to frame the complexities of mobile food vending that is both a social production of encounter and exchange, as well as a contested activity. For example, city governments often limit vendors' freedom in order to support unfettered automobile circulation and restaurant opposition, given their visible economic value. Vendors must, therefore, continuously negotiate urban spaces and argue for their right to occupy the city.

Lefebvre criticized the Marxists' emphasis on economic determinism and sought to explore the everyday realities of urban life. "The Right to the City", originally published in *Le droit á la ville* in 1968 (and republished in English in *Writing on Cities*),[4] has since appeared in numerous works on social theory that attempt to decipher inequalities in the built environment.

Calling for citizens to be active and engaged, his concept of *the right to the city* has been used to frame social justice debates surrounding issues of housing, homelessness, and public space. His concept also expressed his deep concerns with the expansion of capitalism, and the decline of the *oeuvre*. *Oeuvre*, a term suggestive of an opportunity for inhabitation and performance, refers to the "information, symbolism, the imagery and play" in daily life.[5] Lefebvre contends that "the right to the *oeuvre*, to participation and appropriation, are implied in the right to the city". Lefebvre's work demands a renewed right to urban space where "the working class can become the agent, the social carrier, or the support of this realization"[6] as opposed to merely a functional habitat shaped by the needs of power and capital. Lefebvre shows the struggle for food vendors' rights makes urban space a multifaceted social and political product.

This research seeks to develop an accurate description of emerging activities in cities that combines real-time data with more qualitative forms of urban analysis. Using an integrated approach of qualitative and quantitative techniques can demonstrate how online communication creates new spatial and temporal relationships in cities.

Data Collection Methods

Food vendors in Charlotte, NC, were analyzed by closely documenting their daily operations, tweet content, and spatial locations over an extended period of time. Six mobile food vendors, of the 20 regularly active ones in Charlotte at the time, were chosen based on the greatest number of 'followers' (> 1,000) tallied on each vendor's Twitter account and their repeated presence at Charlotte's popular "Food Truck Friday". Six trucks were identified: Tin Kitchen, Papi Queso, Wingzza, Roaming Fork, Herban Legend, and On The Go Cupcakes. Each vendor was interviewed in person and by phone twice on topics including, length of time in the business, operating methods, approach to using Twitter, scheduling procedures for events and locations, feedback from customers, and menu items related to location or time. Each vendor was visited at least three times on-site to record the arrangement of the truck and customers through diagrams, photographs, and time-lapse video; on 15 November 2012, 1,000 tweets were collected from each vendor, which included, the vendor's account name, the date and time of the tweet, the number of retweets, and the tweet content. The tweets spanned periods of four to nine months depending on the vendor's tweeting frequency.

Urban Space Analysis

Four urban spaces were chosen to investigate their relationship to traditional planned gathering places that typically include areas for seating, shade, open space, views, and parking. This research takes into consideration Lefebvre's stance that the *oeuvre*, or playful social urban life created by the citizenry, is suppressed by more powerful actors aimed at protecting social and economic order. Therefore, the focus of our urban space analysis compares the qualities of planned social spaces and food truck locations to understand vendors' affordances in their right to the city. Using direct observation, the immediate site features and social activity occurring at the truck locations (e.g., customers, site amenities, and any other co-located trucks) were documented along with their position within a larger urban context (e.g., nearby land and building use) to understand the types of spaces they frequently inhabit (Illustration 6.1).

Analyzed together, the vendor locations reflect many of Charlotte's geographic and cultural characteristics: a compact downtown core, a decentralized landscape linked by an extensive roadway system, business park developments located on the southern periphery, the university, and major thoroughfares and areas of commerce throughout. These locations embedded throughout the city reveal an uneven spatial and temporal distribution.

The first space is a privately owned, vacant dirt parcel situated on the periphery of Charlotte's central business district, also known as Historic South End. With the help of Charlotte's downtown association, the site became home to "Food Truck Friday" in October 2011. The downtown association mentioned several factors that were considered in determining this location: visibility from pedestrian paths, the scale of the space so that it could accommodate food trucks while also being intimate enough for patrons, proximity to public restrooms, available parking, and few nearby food establishments. This site's context resembles a traditional gathering space with its abundance of open space and proximity to retail and public transit. Yet, the site is left unused otherwise, lacks shade devices or trees, lacks infrastructure for activities or sitting, and has relatively low pedestrian activity other than Friday.

The second location caters to lunchtime patrons multiple days of the week on an active vehicular thoroughfare adjacent to a culinary university and a residential condominium. Key factors in the success of this location are truck visibility to vehicular traffic, accessibility by foot to residential and educational buildings, and nearby infrastructure for seating and shade. However, space does not serve as a designated gathering space, rather the presence of the food trucks parked on a busy street activates nearby spaces.

The third space is a large parking lot located in an office park south of the central business district. Patrons working in nearby offices take food to their cars to eat alone, sometimes in pairs, while others take food to their office building. Given the short lunch hour and the lack of seating options, this location has large influxes of people and then lulls of emptiness leaving patrons little opportunity to socialize. A vendor mentioned that her more dedicated customers order in advance, allowing them to avoid lines. The design elements of this space have no bearing on the vending experience or traditional gathering spaces.

The last location is nestled behind a local brewery in the Historic Arts District northeast of downtown. The paved parking lot accommodates a single truck nightly that is organized by the brewery owner. Given the remote location of the brewery, the majority of the customers drive up; however, customers tend to occupy the site for long periods of time. The site does offer seating for patrons indoors and outside but lacks open space, shade, street visibility, and pedestrian and public transit access.

While the resemblance between vending locations and traditional parks and plazas varies among sites, this investigation shows vending locations were never initially conceived spatially. Despite the lack of accommodations or access to dense pedestrian areas, the locations become active social settings with the presence of a food truck. As shown in Illustration 6.1, they are situated in predominantly privately owned and underutilized land. These spaces are instantly transformed with makeshift seating and vibrant mobile restaurants. While the vendors' locations were not identified as spaces of conflict in a fight for rights to use space, the government's permitted use of private property and the locations in unfamiliar and peripheral public land separate from established business sectors of the city indicate they marginalize vendor activity and limit their opportunities. Vendors' right to the city is not equal to other forms of commerce by virtue of the spaces they are allowed to inhabit. Despite this injustice, the appropriation of urban space by vendors continues to grow and offers options for social life to occur unexpectedly.

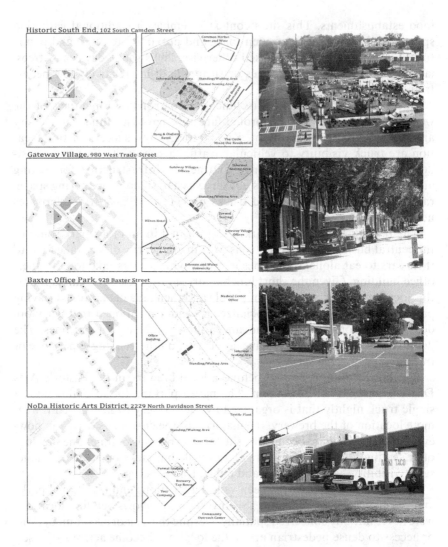

ILLUSTRATION 6.1 Mobile truck locations analyzed as traditional gathering spaces.

Social Media Analysis

The next steps of the analysis involved examining the content of each vendor's 1,000 tweets using a simple automatic topic analysis. This effort provides a set of topics that naturally emerge from grouping verbally similar tweets together. Text analysis of tweets is difficult using traditional automated topic analysis due to the short length of documents, heavy use of slang and abbreviations, and noise from URLs, attached images, and automated tweets from applications. More sophisticated methods such as Latent Semantic Analysis (LSA), did not produce better results with our data, likely because the short length of the documents violates the assumptions of these methods. Therefore, a simple topic analysis method based on k-means clustering was used.

First, the tweet data were simplified by extracting the most frequent 1,000 keywords minus a standard list of *stop words*, or words to ignore (e.g., overly common words such as *the, an, of*), augmented with tweet-specific stop words including *RT, #,* and URL components such as *http* and *t.co* (the standard beginning of a URL automatically shortened by Twitter). These 1,000 keywords become dimensions in a transformed dataset. Instead of being a string of words, each tweet is a vector of zeros (does not contain the keyword) and ones (contains the keyword). This is known as a "bag of words" technique since the order of words is ignored. This results in a dataset in which each tweet is a point in a very high-dimensional space.

K-means clustering was performed using the Euclidean distance between the vectors that represent the tweets. K-means clustering takes several clusters as an argument and classifies data items according to the tightest clusters when space is separated into that number of groups. The analysis with values of K ranging from four to ten clusters were generated and examined the results by hand. Based on the qualitative results of interviews, the topics produced by the eight-cluster analysis were judged to be the most meaningful (Illustration 6.2). The top ten features for each cluster are listed, along with the number of tweets in the dataset, which are classified as part of the cluster. It is possible to derive semantically separable clusters because the tweets are more restricted in content than most collections of tweet data since they come from Twitter accounts with the common goal of marketing a food truck. The largest cluster, labeled "Miscellaneous" is not a clearly defined topic and contains several generic terms found in other clusters. The Miscellaneous cluster is largely an artifact of this off-topic tweeting and as a result, is more affected by the difficulty of meaningfully classifying very short documents based on a bag-of-words approach.

The other seven clusters reveal more meaningful patterns that relate to the trends seen in interviews and the on-site analysis. Tweets about the trucks'

	Miscellaneous	Schedule	Food Truck Trend	Gratitude	Food Truck Friday	Locations	Truck Mentions	Menu Items
Top features	thanks	lunch	food	thank	truck	pecan	@thetinkitchen	sweet
	@wingzzatruck	come	truck	great	food	1111	@herban_legend	potato
	great	see	trucks	sold	friday	ave	@napolitanosmkt	hash
	cupcakes	today	great	see	@southendclt	plaza	@sticksandcones	special
	today	us	@thetinkitchen	us	camden	midwood	@autoburger	w/
	just	st	come	awesome	park	dinner	@southerncake	today
	see	trade	@herban_legend	day	southend	fork	#foodtruckfriday	tacos
	@thetinkitchen	30	#clt	rock	tonight	roaming	@cltfoodtrucks	bacon
	tonight	tomorrow	beer	guys	see	tonight	@papiquesotruck	chorizo
	us	schedule	charlotte	support		8-May	@turkeyand	
Tweets	5201	999	413	300	266	189	111	26

ILLUSTRATION 6.2 Eight topic clusters derived from k-means clustering analysis.

schedule, locations, menu items, and gratitude are the most common. The "Food Truck Trend" cluster, which contains a mix of hashtags, mentions, and words relating to the phrase "Charlotte food trucks", is related to trending topics about food trucks that emerge around popular events. The "Truck Mentions" cluster, which includes vendor usernames preceded by the @ symbol and is used to mention or reply to another Twitter account holder, suggests an active and ongoing dialogue occurs between vendors and their customers. The "Food Truck Friday" cluster contains a series of terms such as rally, Friday, and Southend that relate to topics about the most popular event that takes place in the city. The clusters "Schedule" and "Location" include terms about time and location-based information, which are necessary aspects of businesses reliant on mobility.

After eliminating the "Miscellaneous" cluster, the remaining seven clusters were used as a framework to compare the total tweets and retweets for each vendor. Illustration 6.3 shows that the "Schedule" cluster was tweeted and retweeted most frequently across all the clusters highlighting the importance of advanced planning by vendors. Also, the clusters "Food Truck Trend" and "Food Truck Friday" that include popular general terms about food vending and are consistently retweeted across all vendors shows tweets can frequently travel between Twitter accounts. Furthermore, tweets that include terms about "Gratitude" generated very high retweeting for the vendor @WingzzaTruck and @roamingfork showing positive dialogue and relationships can extend past the point of sale. Similarly, the owner of @roamingfork mentioned, "Rather than focusing on money, I think first about quality and customer service. My customers need to know that they are appreciated and I am thankful for the opportunity to get out there and give them what they expect". Last, the "Truck Mentions" cluster shows a consistent amount of retweeting suggesting a strong back and forth communication exists between vendors and their customers which may look like, "First @WingzzaTruck stop of the year! They are always so nice and the food is delish!"

A two-month timeline was constructed (September to October 2012) showing the days in which food truck events occur, the numbers of event-related tweets, and the number of nonevent-related tweets (Illustration 6.4). The total count of events, event-related tweets, and nonevent-related tweets are identified for each vendors' timeline. In general, the total number of tweets for any one event ranged from one to nine, the time frame ranged from three weeks before the event up to and including during the event, and event-related tweets most frequently occur on the day of an event or within one week prior to an event.

First, graphic analysis reveals that multiple events occur on a single day more frequently during the work week. This challenges the presumption that vending is a leisurely weekend destination. Employees now find food trucks to be a common food choice during lunch hours. Subsequently, the owners of @herban_legend, @TheTINKitchen, and @roamingforkNC mentioned

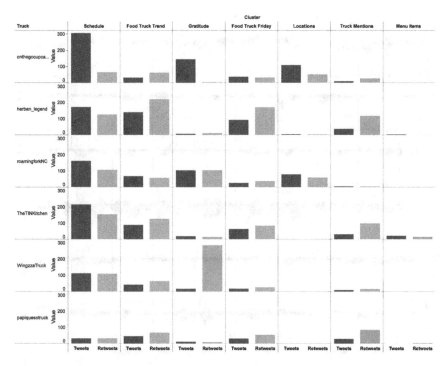

ILLUSTRATION 6.3 Total tweets and retweets organized by vendor and cluster topic.

in interviews that they serve lunch three days a week at remote business park locations or college campuses.

Second, the nonevent-related tweeting reveals vendors often communicate about a variety of aspects, such as types of food, customer feedback, truck-operating issues, and their personal life. A tweet by @roamingforNC states, "In Costa Rica ... Rejuvenating ... Relaxing ... What's happening where you are?" The vendor @WingzzaTruck tweeted, "Hey #TeamWingzza if you could add one item to our menu what would it be? Working on some things for you! #CLTFood". This tweeting proves to be a consistent type of communication throughout the entire two-month mapping for all vendors, suggesting that it is essential to continue dialogue and engage customers in ways that are personal and positive.

Third, some vendors when compared show an inverse relationship between the number of event-related tweets and the number of events (see Illustration 6.4). For @herban_legend, who has a total of 68 events and 95 event-related tweets, and @papiquesotruck, who has 41 events and 163 event-related tweets, the number of tweets bears no relationship to an increase of events. Interestingly, maybe this is related to @papiquesotruck's later opening in 2012 and their need to attract more followers. The vendor

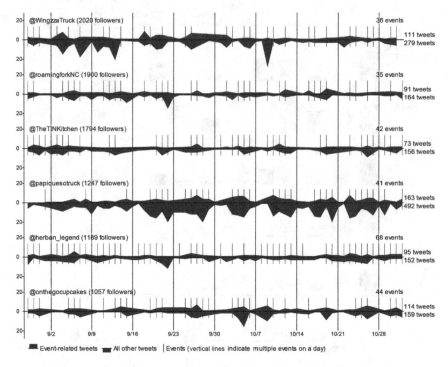

ILLUSTRATION 6.4 A two-month tweet timeline of each vendor.

@papiquesotruck mentioned in an interview, "My website doesn't drive my business, social media does. I can tell that my customers use Twitter a lot because we will post a secret menu occasionally and within minutes we'll have people standing in line".

Using the vendors' event-related tweets, locations were identified by the frequency of mention and tagged as either "one-time" or "repeated". Most locations were named multiple times for a given vendor (e.g., Food Truck Friday) while others were only mentioned once (e.g., Democratic National Convention). Then the number of each type of event was identified and aggregated to determine how many days in advance vendors tweeted (Illustration 6.5). One-time events are tweeted about more often and earlier in advance (2.61 times on average; 1.84 days in advance) than repeated events (1.84 times on average; 0.86 days in advance), although they take place less often. This suggests there is high value in online communication when familiarizing customers with new locations. The mobile nature of their practice can serve as a disadvantage compared to restaurants that have predictable and fixed locations. The vendor @papiquesotruck is well aware of this challenge stating, "I like building long-term partnerships with businesses so that my customers know where to find me on a regular basis".

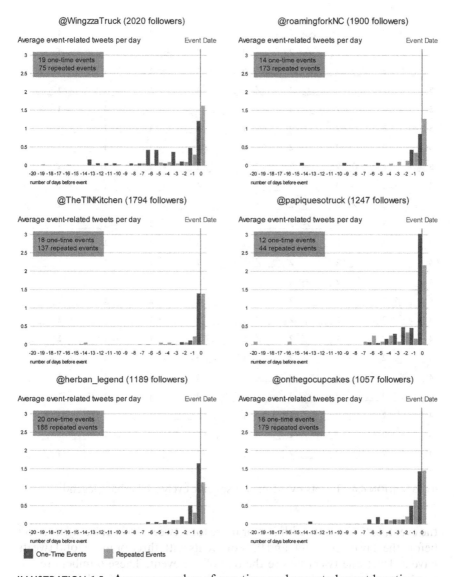

ILLUSTRATION 6.5 Average number of one-time and repeated event locations.

Next, a single vendor was investigated, @herban_legend, to better understand event-related tweeting concerning their events. A sparkline was generated for each of the 209 total events that show corresponding tweets identified with each event (Illustration 6.6). The largest number of tweets related to a specific event is nine and a pattern of tweeting emerges well in advance of the day of the event and often multiple times. This asynchronous relationship between the event and online communication is represented in the following: 33% of all events have a single tweet at least one day before

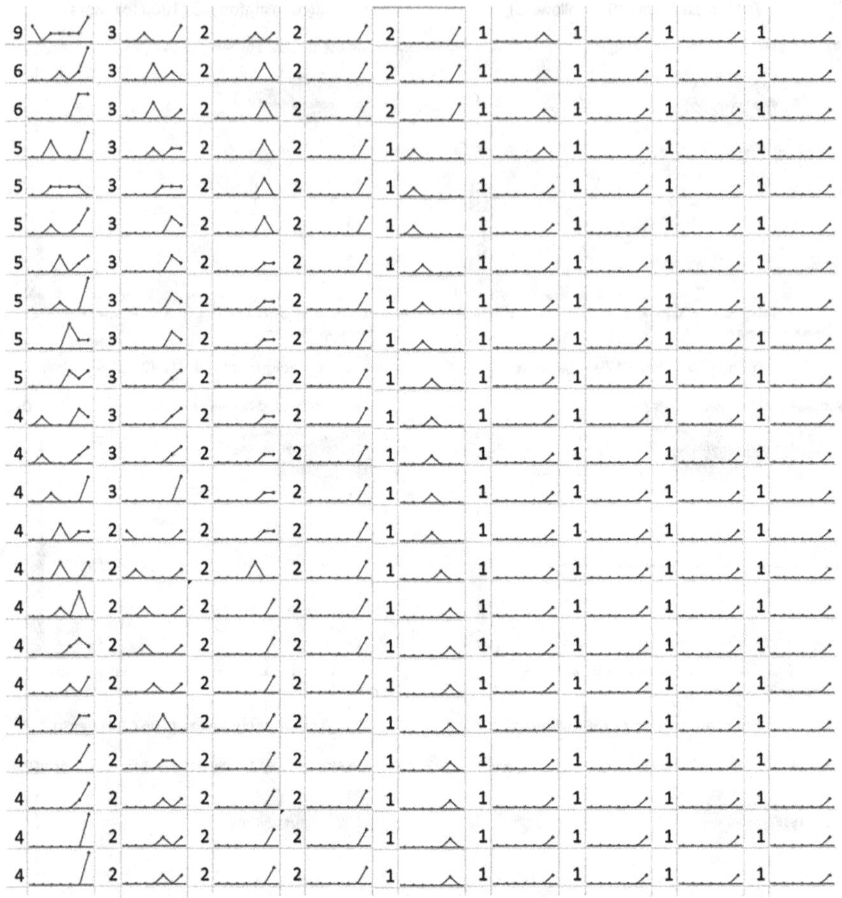

ILLUSTRATION 6.6 Tweets over time for specific events @herban_legend.

the event, 61% of events with two or more total tweets have at least one tweet before the day of the event, and 92% of events with three or more total tweets have at least one tweet before the day of the event. These numbers reveal vendors' routine practice of planning and announcing events ahead of time. Unlike older vending populations, these vendors can construct an event and ensure a predictable customer base well before it physically occurs.

Mobility, Social Media, and Urban Space

This research examined mobile food vending in the United States as a form of urban spatial occupation linked to online communication. While urban space remains physically stable amid new forms of ephemeral activity, the dialogue afforded by information technology further extends the reach of an audience beyond the physical propinquity of a food vendor and creates a demand for new food experiences in unfamiliar urban spaces.

Presumably. mobile food vendors might find places in the city that closely follow and reinforce the planned spatial pattern located on primary boulevards and parks and open spaces. If this were true, existing methods from urban design would be sufficient to explain and plan for their growth. On the other hand, the locational decisions of a truck might reflect the aggregated demand derived from information technology, shifting locations in real-time to meet the greatest number of customers. The truck could move many times over the course of a day, a "just-in-time" restaurant. In this scenario, traditional urban space methods are irrelevant; the trucks would move frequently and without regard to the nature of the space they occupy. The study showed the local spatial arrangement of the trucks across the city was highly variable. Sometimes, the trucks reinforce a more or less predictable arrangement of urban form (e.g., Food Truck Friday), but often the spatial arrangement responds only to the location of potential customers (e.g. office parks and university campuses).

The predominant use of private property and unfamiliar, peripheral public land by vendors indicates where vendors find most opportunity. It also shows that locations where they are permitted to vend avoid the central business district where brick and mortar restaurants fight to keep vendors away. Vendors' rights to occupy urban space are not equal to other forms of food commerce by virtue of the spaces across the city that they are allowed to inhabit. Despite this spatial injustice, the appropriation of urban space by vendors continues to grow and offers options for social life to occur in an unexpected way.

This study also revealed that data, space, and time form a tightly woven network that functions before each food truck event. Mobile food truck locations were announced several times in advance of the scheduled date, creating an audience for a particular location at a particular time. This allows the vendor the flexibility to reach more than a spatially adjacent audience, as opposed to prior modes that relied on the repeated presence or word-of-mouth communication. This unique condition reveals a dynamic found in Latour's ANT that explains how physical, social, and technological networks are integrated and operate heterogeneously to sustain emerging innovations. Subjects (i.e. food truck vendors) and objects (i.e. social media and urban space) undergo a series of translations or transformations to account for the unpredictability of where vendors can locate on any given day. The insecurity of having a sufficient customer base in an unknown location creates a mutual relationship between vendors and social media.

Overall, the research shows that neither data nor urban space considered separately could adequately explain vendors' behavior. Existing descriptions of space using canonical methods of urban design could not describe the locations of the trucks with any degree of certainty, while at the same time, investigations of the data network using either topic modelling or temporal information were only meaningful with some knowledge of the spatial locations. The interdependent relationship between data, space, and time generates new forms of social activity in unfamiliar settings.

The emphasis on the spatial over the temporal and digital in urban design has become a serious drawback, making the understanding of new forms of urbanism difficult to see or analyze. Considering only the informational network would be equally one-sided. By using mixed methods, this research proposes hybrid forms of urban analysis that are informed both by a qualitative recognition of human urban places and a quantitative understanding of large flows of data.

Notes

1 An early version of this chapter was published as Wessel, Ginette, Caroline Ziemkiewicz, & Eric Sauda. (2016). "Revaluating urban space through tweets: An analysis of Twitter-based mobile food vendors and online communication." *New Media & Society, 18*(8), 1636–1656.
2 Brennan, A. (2014). *ISIS World Industry Report 0D4322-Food trucks*. Retrieved from http://clients1.ibisworld.com/reports/us/industry/default.aspx?entid=4322
3 Portillo, C. (2014). "Charlotte food truck vendors criticize proposed regulations." *Charlotte Observer*, 21 April.
4 Lefebvre, Henri. (1996). *Writings on Cities*, trans. Eleonore Kofman & Elizabeth Lebas (Eds.), Oxford, UK: Blackwell.
5 Lefebvre, Henri. (1991b). *Critique of Everyday Life: Volume*, trans. J. Moore, London: Verso [first published, 1947].
6 Lefebvre, Henri. (2003). *The Urban Revolution*, trans. R. Bononno, London, UK: University of Minnesota Press, 1–22 [first published, 1970].

7

IRANIAN VENDORS

with Guest Author Neda Kardooni

Cities and urban spaces are a critical arena where government policies and human lives interact. In contemporary Iran, cities are, as Asaf Bayat puts it, where poor people whose distinguishing factor is their informal economic activities make "slow progress". This poverty, however, is not solely financial. Many disenfranchised groups such as women, ethnic groups, the elderly, and LGBTQ for various reasons face inequalities and injustices in terms of policy restrictions and the potential for progress. These populations occupy and live in urban spaces not only to improve their economic situation but also to find a space to achieve justice and equity through increasing their presence in the public. For them, increasing presence in public space is synonymous with achieving a broader set of opportunities where the "right to the city" translates into many other rights such as freedom of speech and choice. In a sense, cities are paradoxical spaces where the disenfranchised are deprived of their rights but they also fight for justice, highlight their differences, and reflect their legitimate expectations.

Urban spaces in Iran, especially after the 1979 revolution, are a bedrock of such social changes and progress that are not represented in official government policies.[1] Poor populations, ignoring official government policies, often occupy public spaces not sanctioned for economic activity to engage in commerce and economic activity.[2] Street vendors, for example, find areas in cities that have a high pedestrian traffic flow such as public squares or even metro cars to sell different kinds of goods to pedestrians. However, even these activities that are against official policies are not equally and equitably available to everyone. Due to harsher governmental rule and cultural customs, language barriers, ethnic customs, and lack of proper infrastructures, systematically disenfranchised people such as women, ethnic minorities, individuals with disabilities, and the LGBTQ have less potential to engage in street vending.[3]

DOI: 10.4324/9781003026068-9

The growth of social media platforms in Iran is slowly changing the dynamics between people and governmental and institutional norms. The disenfranchised populations in Iran use social media platforms to engage in activities that might face restrictions in physical urban spaces. Typical of such activities are a myriad of women-owned or minority-owned social media pages that sell goods by directly connecting to customers, as well as the ethnic minorities of Iran who try connecting to a wider customer base. These activities are not completely separate from urban spaces. In a sense, disenfranchised populations in Iran are engaged in "urbanizing and spatializing social media" by constantly switching between social media and urban space, or engaging in online activities that recall the look, feel, and even the sound of vendors in urban space.

The aim of this chapter is to study how both social media and urban space are used separately or together by disenfranchised populations to engage in informal commerce and as Asef Bayat would describe, *"The Quiet Encroachment of the Ordinary"*. In the first section of this chapter, a brief overview of street vending in Iran is provided. The main resource for this section is Asef Bayat's book on the same topic titled *Street Politics, Poor people's movements in Iran,* as well as some of his more recent publications. In his book, Bayat describes how poor populations in Iran reclaim urban spaces through the act of street vending as a form of political and economic dissent. More recent information from written and online sources are included to make the arguments for contemporary cities.

In the next section, a brief history and an overview of social media usage in Iran are provided which brings together current information about Iran's social media landscape with the personal experiences of the authors. Next, through a series of interviews with Instagram business owners that engage in informal commerce and topic modelling of commerce-oriented channels, a detailed view evolves of how this historically street-based phenomenon is transformed through social media. Finally, through a series of anecdotes and examples, a discussion is presented on how urban and spatial behaviors and experiences impact informal social media in e-commerce.

Street Politics and New Ways to Reclaim the Right to the City

The dynamics of informal economies in most countries around the world are very complex with deep roots in the political, cultural, ethnic, and economic fabrics of these societies. Iran is a clear example of how poor, and even middle-class populations reclaim public space and engage in economic activities not sanctioned by local and national governments. The complex dynamics of Iranian street commercial activities have been carefully analyzed by Asef Bayat. His claims street vending in Iran and many other parts of the world is in fact a form of political dissent against governments manifested in urban spaces. He describes street vending as a way for poor populations to "open up new spaces within which the poor can reclaim their right to the city"[4] and engage in social non-movements that are "collective actions of non-collective actors; they embody shared practices of large numbers of ordinary people whose fragmented but similar activities trigger much social change".[5]

Iran is a country in the Middle East with a current population of approximately 80 million. Similar to many developing countries, Iran's population is rapidly urbanizing and currently more than 75% of the population lives in urban areas. Besides the official language of Farsi, many different ethnicities with different languages and cultures live in different parts of Iran including but not limited to Arabs, Kurds, Bakhtiaris, Balouches, Lurs, and Azeris (Turks). The majority of Iran's population are Shia Muslims, while there is a considerable Sunni population in some regions, as well as followers of other religions including but not limited to Christianity, Judaism, and Zarathustrianism.

The official religion of Iran's government is the Shia branch of Islam which has impacted Iran's urbanization and formation of urban spaces before and after the 1979 revolution. The Islamic Republic of Iran was formed after a revolution in 1979, in which people of different factions overthrew the last Shah of the Pahlavi Dynasty. The following years are full of domestic and international incidents such as the US embassy hostage situation, a number of cultural transformation and political conflicts, and an eight-year war that shaped Iranian cities. Throughout all this turmoil, people have engaged in political activities in the public spaces of every Iranian city.

Iran's dense and crowded streets are full of life and teeming with diverse activities. Aside from major political events, Iran's poor and disenfranchised populations engage in an ongoing act of *de facto* political anti-government behavior through continuously creating economic opportunities in urban spaces in the form of street vending. Bayat describes these activities as "quiet encroachment of the ordinary, a silent, patient, protracted, and pervasive advancement of ordinary people on the propertied and powerful in order to survive hardships and better their lives".[6] By moving through and setting up shop in crowded streets, shopping centers, and even subway cars, street vendors expose and sell their goods to a large passing population (See Illustration 1). Bayat describes two aims for these populations, (1) redistribution of social goods and opportunities, and (2) attaining autonomy. He also notes that it is important to remember that informality is likely not a preference for the poor populations, it is a necessity brought about as an alternative to the constraints of formal structures.

Since these urban activities are not sanctioned by the government in Iran, there is a constant struggle between these poor populations and the government. When the cumulative growth of "encroachers" is moved beyond tolerable points, the state powers try to erase these activities from urban spaces. However, often the government's crackdowns are not completely successful, since the vendors have gained visibility and move around freely and fluidly. Under these constant struggles between the policy enforcers and the disenfranchised populations, street vendors often are successful in improving their situations, albeit sometimes marginally.

ILLUSTRATION 7.1 Street vending in Iran in Metro car, Produce market, side of the street.

The streets of Iran are not always equally and equitably available even for informal street vending. Muslim straight men have a dominant presence in public spaces. Due to Islamic regulations and cultural limitations, women sometimes have more difficulties in engaging in street vending. These limitations do not necessarily prevent the more disenfranchised populations such as women from engaging in such activities. For example, there are moving subway cars in Tehran where women are constantly engaging in informal vending (See Illustration 7.1 top left image). In cities in the north or the south of Iran where the ethnic cultures are less limiting to women, there is a more prevalent presence of women vendors (See Illustration 7.1 top right image). Religious minorities and women also engage in their own specific struggles through informal vending. The question of how social media might enable disenfranchised populations to engage in informal economies requires an understanding of the use and restraints on social media in Iran.

Social Media in Iran

The Internet became widely accessible in Iran at a rapid pace and almost simultaneously with the rest of the world. Reports show that 60–70% of Iranians are internet users.[7] Many of the prominent social media sites such as Facebook, YouTube, or Twitter are officially banned or "filtered". There are many services and social media sites that are accessible to Iranians including government-approved and sanctioned websites that serve official internet activities such as buying and selling goods or approved social networking. There are also many replicas of western apps such as Spotify or YouTube clones that are available and adhere to the limiting rules of the Iranian government.[8]

However, downloaded smartphone apps in Iran show that the worlds' most popular social media platforms are frequently used in Iran.[9] Telegram, Instagram, and WhatsApp each have more than 45 million users in Iran, while other domestic or international social media platforms see daily users numbering in millions.[10] There is a constant struggle between social media users and the official government of Iran. These social media platforms are used daily for activities such as secure communications and have been instrumental in various political protests.[11] Due to their influential nature, the government of Iran in numerous instances attempted to block and filter many of these platforms. Twitter, Telegram, Facebook, and YouTube are among the banned applications in Iran. To defeat these restrictions, users use filter bypassing technologies such as VPNs to access these media.

The usage of social media platforms in Iran goes beyond political dissent and is embedded within various fabrics of society. One of the most prevalent social media usages in Iran is for commercial activities. Although there are dedicated and sanctioned websites for engaging in commerce, a large number of businesses are created in Telegram channels or Instagram pages. One reason for a lack of interest in official online businesses is the relative ease and inexpensiveness of these platforms, as well as the much more relaxed official surveillance and enforcement of laws and cultural norms.

Informal Commerce and Social Media

Economic activities on social media platforms such as Telegram and Instagram are used to highlight the goods they sell, connect with customers, and receive orders. According to Iran's government representatives, 400,000 pages on Instagram are currently active with more than 5000 followers and about 28% of those are commercial entities. About 48,000 of those Instagram pages engage in informal and semi-formal economic activities.[12] The dynamic between people engaged in informal commerce and governments on social media also manifests in street vending. Similar to how urban spaces are restricted and street vendors fluidly migrate from one location to another, the explosive growth of informal commerce on social media started with the direct messaging/social media platform Telegram which is not routinely used for commerce. However, for various political reasons, Telegram was blocked and filtered by Iranian authorities. The next virtual location for informal commerce was Instagram. In 2018, Iran's executive branch including the president and his minister of information cited the importance of Instagram as a source of income for users, and the economic devastation caused by a recent blocking of Telegram.[13]

Additionally, an advertisement channel on Telegram with more than 30000 followers provides some examples of such informal commercial pages on Telegram and Instagram. Out of 20 topics, the five most prevalent topics in a topic model using Nonnegative-Matrix-Factorization (NMF) of approximately 1400 posts showed commerce-related terms such as blouse, dresses, best-fit, sizes, great cut, clothing, etc. Images in these posts show that many of these businesses focus on women's attire since in physical locations there is a tighter enforcement of Islamic attire (See Illustration 7.2). To better elaborate on these differences, compare the images of women selling clothes in a metro cart in Iran (Illustration 7.1) to the images found on social media spaces.

ILLUSTRATION 7.2 Instagram posts showing clothing not following Iran's legal and cultural norms.

Interviews and Thematic Analysis

To better understand how Instagram is connected to urban space in the informal economy, a number of commerce-related Instagram pages were contacted for a remote interview. A majority of the individuals who were contacted were not comfortable being interviewed about their businesses due to fears of retribution by the government. However, 13 out of 31 pages responded, and consequently, were interviewed. Ten of the interviewees self-identified as female, and the other three as male. The pages covered a wide variety of business types that sold or designed flowers, cooking and food, music lessons, handicraft arts, books, and clothing. All of the interviews were conducted in Farsi. The interviews were conducted in a semi-structured manner and consisted of the interviewee's motivations for using social media, the kinds of social media accounts they use, specific ways they use each social media platform, whether they conduct business in a physical space, and finally, how social media connects them to public spaces and their customers. Notes taken during the interviews were digitized, translated, and analyzed using thematic analysis.

This group of businesses cited passion toward the job, the very limited marginal cost of starting a social media business, and ease of starting a business on social media as their rationale. For example, a female florist who sold flowers on Instagram said that "I have been interested in flower design since I was a child". She mentioned that she held a Master's in industrial engineering but this page helped her business flourish. Later in the interview, she said:

> As a woman, personally, I do not have any problem with being in society because I am a very social and extroverted person and I like to be among people. The only reason that I chose social media for my business was to save on initial costs.

A female porcelain designer referred to the ease and freedom to develop her ideas by mentioning that "social media provides a kind of limitless freedom for developing my ideas and art". Similarly, she described how social media helped her start a business in her own house and minimize the costs to only required tools for porcelain production:

> Although I started this business in a very small workshop in a part of our house, I needed a relatively large initial cost to buy the necessary tools, which my father helped me. I needed a pottery kiln and paints, glazes and other tools.

Another group of interviewees explicitly mentioned financial hardships and poverty as their reason for starting a social media business. A female piano teacher who provides lessons to her students through Instagram said that "I started to work through social media, because I have financial problems in my life". Similarly, the owner of a cooking and baking page who identified as female, mentioned that "the main reason is my financial issues, and the second reason is that I am a perfect cook, so I decided to start a page so I could accept orders for catering". Another female page owner, who sells used branded clothes, bags, and shoes discussed how losing her job motivated her to start a business on social media:

> It's been a while since I lost my previous job and I thought it would be better to start a business for myself. As I have all of the time in the world, so using social media I can manage my time better and be more efficient.

Six of the interviewees stated that social media increases the number of their customers in comparison to only having physical locations. A female business owner who makes and sells organic granola bars on Instagram argued that, "[social media] has increased my customers and I can sell more granola bars through social media". A male student, who, on Instagram, sells second-hand books on the street in a major bookselling zone in Tehran also said that "social media has increased the number of customers I have had".

Furthermore, many of the interviewees referred to mass and direct communication affordances of social media as another primary benefit to them. They discussed the different ways they use social media platforms to improve their business. A male artist who creates wood artifacts discussed how he uses "Instagram to both communicate with customers and also receive their feedback and comments". The female piano teacher described how due to the massive size of Tehran, living in the east side makes it difficult to connect to customers in other parts of the city physically: "I live in the east of Tehran and it is very difficult for me to access the west or the south, so I can communicate with all of my student and trainees through my page". Interviewees also described using social media as a means for advertisement. As an example, the granola maker discussed how she communicates with close friends or customers to advertise their page to others: "We get help from our friends, family to advertise our product to their friends".

Due to the remote nature of holding a business on social media platforms, some of the interviewees described creative ways they use other affordances of social media. The female cook who bakes and sells cakes on social media described how she bakes live on Instagram to help increase trust in her products:

> People prefer to see how I make one cake. So sometimes I set up a live stream and I share all of the process with my customers. So they can see which materials I use, and how I make a cake from 0 to 100.

Later in the interview, she described another strategy for connecting with her customers on social media: "I record the birthdays of all the customers and sometimes I send them cakes as a surprise and as a gift for their birthday". The piano teacher utilized another interesting advertisement strategy:

> I created another page for my students, and to encourage those who have played the piano well, I post their video (the progress of my students) on that page. This will both encourage my students and make more people trust my work.

Besides the minimal costs associated with transforming their residence to a business through social media, some of the interviewees mentioned that social media is a good starting point for them and that they would need a fixed location if their business grows. As an example, the woman who sells granola bars on social media said, "If we have more sales, we definitely need more facilities, which are not possible at home, and we must consider a space in the city". The jewelry designer mentioned how she uses locations temporarily as a way to create a larger impact for her work: "Sometimes I present my work in a cultural center and my work is well received, but renting a small booth in the cultural center is very expensive". The interviewees also discussed greater access to their customers through social media. For example, the street vendor who sells used books both on the street and on social media mentioned how people from other parts of the country purchase his goods: "through [social media] I have customers from all over Iran, not just Tehran". Later in the interview, he mentioned how he connects customers to other physical locations through his social media page: "My brother's partner has a bookstore on Enghelab Street. I advertise for them as a vendor and send customers for them".

There were also discussions of social media providing fewer restrictions for women in comparison to physical urban spaces. To this point, the flower designer discussed how social media could feel safer for women: "I think there may be women who have chosen to conduct business through social media because they have restrictions on being in the city and they feel more comfortable and safe when they use social media for their business". The porcelain designer mentioned that physical spaces do not provide the level of freedom that she needs: "I think being only in the urban physical space does not have this opportunity". Another example includes a mother who designed and sold accessories as a source of income. She mentioned that "I do not want to work in an urban environment until next 10 years, because I have small children at home and I have to take care of them as well".

Even with all the benefits, attractiveness, and unique affordances of social media, the majority of interviewees mentioned that at some point they would be very interested to have a permanent location for their business. A male online clothing page owner mentioned that he prefers to be part of a community: "Even if my income is great through social media, I still like to have boutiques in the urban space. Because I like being in the community.

It gives me a sense of life". Similarly, the piano teacher mentioned that: "Despite being an introverted person, I prefer to have a permanent place in the city and follow my activities on social media". The lady who sells second-hand clothing items on social media, discussed how her business is better suited for physical spaces and her customers also think so as well:

> I have plans to rent a small shop in the city, because these types of products need to be seen, checked and even tried before buying it. Many of my customers ask me if they can wear the clothes before buying them.

The interviews highlight key findings of social media as a medium for informal economies. Interviewees mentioned economic needs and poverty and greater freedom to disenfranchised populations such as women. The finding that most interviewees either operated in physical spaces or were interested in doing so in the future raises a question concerning the relationship between physical space and social media.

In order to answer these questions, the research team engaged in an online ethnography process with Instagram business owners acting as customers. Two questions were sent to a large number of Instagram pages in the form of direct messages. The questions were (1) do you have a store or place that I can come see/try what you sell? And (2) can I come and see your goods in person at a specific time? Four groups of businesses were found and contacted using hashtag searches. To understand differences between accounts that operate solely on social media and ones that are hybrid businesses, for each group, unique descriptions of one page that solely operates on social media and one that operates as a hybrid of urban/social media space were provided.

One group of Instagram accounts was identified as **street vendors**. These accounts were found on Instagram using the #دستفروش (dastforoosh) and #بساطی (basati) which both translate as street vendors. Many of these accounts use such terms in their account title or descriptions of their posts. Out of 49 accounts who responded to the research team's queries, 17 responded that they have temporary places in the city and that the customers could go and see the goods they sell. Observing the pages of these accounts shows that some are established businesses that also operate on social media, while others are operate by individuals possibly within their homes. Two accounts that sell women's headwear highlight this significant difference.

For example, one account responded that they operate from a permanent place where customers can see their products and buy in person but they also ship their products to customers around Iran. The account has 17.9K followers and sells shawls, which are headwear that women in Iran are required to loosely wear over their heads. The description of the account starts with a contact number, then reads "shipping to all locations in Iran, single sale or wholesale, hours of operation 6 to 10 pm". The page mostly posts professionally edited pictures of models posing with the shawls. There's also close-up pictures of fabrics used for making the shawls, as well as a couple of pictures of a man who appears to be the owner of the page. The description under most posts contains a similar message; the price of the shawl, a flat price for shipping to any city in Iran, as well as descriptions of the quality and color. At the end of each description, there is a sentence that reads "in order to buy in person, visit [redacted] address". The address seems to be a permanent location close to a bank in a northern Iranian city. The account also has highlighted stories that read "satisfaction of the customers". The page does not seem to represent an actual street vendor, but rather is a shop that operates on social media to reach a wider customer base.

A second account also identifies as a street vendor. This page with 156 followers responded that they do not have a temporary or permanent place. They also said that they solely sell their products to customers through shipping. The account's description reads: "women's products, for ordering please send a direct message, shipping only in Shiraz [A city in Iran]". The pictures do not seem to be professionally edited and only show close-up views of products. In comparison to the previous account, this page contains a much larger variety of products such as makeup, underwear, fabrics, barbies, toys, wallets, and phone cases. It appears that this account is reselling these products, and over time, the nature of the products they sell are subject to change. In terms of the nature of the products, this account closely resembles that of an actual street vendor who sells a variety of objects for a profit.

The second group of Instagram accounts focused on selling **art products**. All of these accounts included the Farsi translations of art, handcraft arts, or creative works in their titles or descriptions. Out of a total of 35 contacted accounts, 10 operated from a physical location. The accounts sell a variety of products ranging from baked goods, pottery and mugs, paintings, and anime and manga arts. Many of the pages show a purely commercial orientation and only focus on selling products. On the other hand, some pages are creative outlets for the artists who also do commission work, or sell their creative work through these pages. Descriptions of two of such pages can better exemplify these differences.

The first page focuses on handcraft arts. The vendor operates mostly on social media with 1,100 followers but also often sells in galleries or temporary places. The description of the page reads: "Handcraft arts shop, here we sell beautiful art for your houses, to order please order using direct messages". The description continues by providing the website for the shop that contains information about the name and picture of the owner who is a young male, contact information including email, phone number, and a fax number. In the about us section, there is a sentence that describes how the owner has sold work in various galleries across Tehran. The account posts professionally edited pictures of different home art products including flower pots, teapots, mugs, and purely decorative objects. All pictures are watermarked by the logo of the page. Descriptions of each product describe the dimensions, price, and material details of each art piece, as well as details of shipping prices.

The second page is an Iranian manga artist who responded that they do not have a physical location. The page has 2900 followers with a description in English that reads: "Anime artist, Iranian, commissions open". The pictures are mostly color or black and white anime and manga sketches. The descriptions of images are each unique to the image, which describes the artist's feelings about the character they have drawn. Some of the posts are videos of the artist's drawing and coloring process. Overall, there is very little information about how the artist sells these anime drawings, besides one highlighted story that shows a video of a drawing in process with an overlaid typography that reads: "commission in process". In comparison to the other art-related page, this page is primarily the creative outlet of the artist but also serves as a way for the artist to make money.

The third group that was queried on Instagram was a group of accounts that used the word or hashtag #تهلنجی (Tahlenji) within their names. The literal translation of Tahlenji is "bottom of boat". It refers to an informal bazaar with stores that sell high-quality goods such as snacks, house products, foods, and small appliances that are smuggled from foreign countries. The story behind the name is that the goods are hidden at the bottom of a boat and have not officially gone through customs at the borders and, thus, are normally less expensive than the officially imported products. In Abadan, a southern border city, adjacent to the border of Iran and Iraq, there exists a famous bazaar named Tahlenji Bazaar that is famous for selling such products. Out of 74 accounts that were named after this famous bazaar in Abadan, 25 responded that they also have stores. The rest of the pages responded that they only ship products to customers. Descriptions of two of these Tahlenji accounts exemplify the differences between those pages that operate on a hybrid area of urban space and social media and ones that operate solely on social media.

The title of the first account translates as "the sailor tahlenji supermarket". Without inquiring from the owner, it is evident that this account also has a store that sells its products. The description of the page reads: "Prices just like Dubai, wholesale and small quantities, shipping to all parts of Iran". The description continues with the physical address and the phone number of the store. The page has 10.7K followers, and also has a link to a Telegram account for direct inquiries. The posts are all of the products that are sold by the store. The majority of the images show at least two of the same products which seem to convey that there are plenty of products available. The pictures are taken in the store and most images' background is a blurred angle of shelves full of products. The description under each image is very generic and includes price, name, and some detailed specifications. At the end of the description, it says "for shipping please direct message us".

The second page also uses the name Tahlenji within the name and the description. However, the owner responded that they do not own a physical location and only ship products to customers. The page only has 226 followers, and its descriptions read: "Foreign food and washing products, spices from abadan, shipping to all parts of Iran". There are no phone numbers or physical addresses present on the page. The images mostly show similar products to the previous store discussed above. However, instead of a store's background, the images are taken over a simple cloth, with a plain background. Many of the images include a single product, while some of them include two objects. The descriptions of the images include specific sentences about the products that are such as "delicious milk chocolate you are going to love" and "extra hot sauce, be careful".

These inquiries show that social media can serve as a space for both established stores with physical locations and individuals who engage in informal commerce. Established stores navigate customers between their physical locations and social media pages. Social media helps increase the visibility of products, connect more customers, and increase physical access to their goods. For individuals without an established business, the story is different. They are primarily using their homes to conduct business. They use the vast opportunities of social media to start informal businesses without the need to go through any bureaucracy or formal process. For many, social media platforms serve as a new extension to the streets of the city where selling goods, despite occasional clashes with the city and government officials over permits, comes as an antidote to the limits and grievances of the poor and struggling populations.

Urban Space and Social Media Commerce

Interviews with social media business holders and explorations of various kinds of businesses showed that social media is serving as a new space for informal commerce, often completely removing the need for individuals to access the more culturally, legally, and physically limiting streets. But are these spaces completely divorced from urban spaces? Reviewing social media pages reveals several examples where businesses use words and phrases as metaphors that communicate experiences directly related to physical locations.

#Tahlenji: Searching Instagram for #Tahlenji, more than 6000 posts show up. It is clear that many pages have no physical association to the very established and famous urban space called the "Tahlenji Bazaar" that currently exists in the harbor cities of Iran (See Illustration 7.3). However, the individuals are interested in the quality and experience associated with going to "Tahlenji Bazaar". In other words, by using the name, these pages become spatial metaphors and allow Iranians to immediately recognize and expect certain products, price ranges, and quality levels offered by a business.

#Friday Bazaar: In Iran, Fridays are the weekend. This hashtag (#جمعه_بازار) with more than 30,000 search results is not related to a specific location with a specific quality such as Tahlenji but weekends when in almost every city in Iran, a specific location is dedicated to the collective

ILLUSTRATION 7.3 Tahlenji Bazaar where a variety of goods are imported and sold.

location of street vendors and artists selling various kinds of products. On Fridays a large number of customers go to these locations and engage with businesses. Posts accompanied by this hashtag are mostly related to small decorative objects, handcraft arts, and inexpensive clothing objects.

#Parvane Parking Lot: Another example of a specific known area serving as a metaphor for communicating the quality and kinds of products is پارکینگ_پروانه# (Parvane Parking Lot) with more than 17,000 search results. This parking lot is the fixed and sanctioned location in Tehran where every Friday street vendors who are primarily students gather and sell various kinds of decorative, artistic, and handcraft arts (see Illustration 7.4). A vast majority of posts coming with this hashtag are close-ups of decorative products that are sold solely on social media.

#Street vending: Some spatial metaphors are not related to specific locations or times, but specific activities in urban spaces. Searching the hashtag "street vending" (بساطی# دستفروش#) yields more than 30,000 search results, many of which are related to posts about rights of actual street vendors. However, almost equally prevalent are commerce-related posts that use this hashtag as a spatial metaphor to possibly communicate temporary, inexpensive prices and a variety of goods.

#Men_are_not_allowed: Another interesting metaphor, was the use of the term ورود_اقایان_ممنوع which translates to men are not allowed. These names or hashtags refer to specific stores that only sell women's clothes and underwear. Due to government restrictions, these stores do not allow the entry of men. Going to these stores, big red no entry signs with a text reading that men are not allowed is a usual sight. Although in social media, there is no actual way to limit the entrance of men, usage of such terms allows surfers to easily know that this page is a business that focuses on women's underwear and clothes.

ILLUSTRATION 7.4 Ad hoc street market at Parvane parking lot.

#we wiil buuuy: A very surprising and interesting example of spatial metaphors is a hashtag that does not relate to any visual experience but an auditory one. Many Iranian citizens have experienced sounds of pickup trucks that slowly pass through every street or alley and with a voice amplifier, yell: "broken refrigerators we will buy, old air conditioners we will buy, used plastic, we will buy,...". This repetitive, protracted sound that emphasizes the last part of the sentence is associated with second-hand, used item purchasers and resellers. This hashtag, which clearly communicates the auditory experience of those sellers, is often used in the description of the pages to refer to the property of the social media page.

A Quiet "Digital" Encroachment of the Ordinary

Prior to the prevalence of social media, informal commerce was conducted exclusively in public spaces. Poor populations create ad hoc political and economic occupations of public spaces. Governments of various scales often clash with and try to dismantle or control these movements. Even though this dynamic is empowering to poor populations, the affordances of public spaces are limited both in terms of equitable access to all populations and the reach to a large number of customers.

Social media has offered a new space for commercial activities. Exploration of social media commercial activities in Iran shows that both established businesses with physical stores and individuals without any commercial presence in public spaces use social media platforms to conduct business. Interviews with Iranian social media business owners showed that social media provides many benefits such as relative anonymity, more freedom, fewer expenses, and a wider geographic reach to a larger audience. A survey of social media businesses showed that businesses with physical locations create informal business pages to take advantage of the affordances of social media platforms. However, these established businesses often explicitly (through their physical addresses) or implicitly (through images that include the store as a background) show the existence of their physical locations and direct people from social media to their stores.

For many people, social media serves as an alternative to streets for conducting informal commercial activities. Women, artists, or many individuals without enough financial power choose social media pages as an alternative to setting up temporary shops in the streets or designated places in their cities. To them, social media offers a low-cost starting point for their business, a wider geographic reach, and a potentially larger audience. Additionally, for some disenfranchised populations, social media businesses offer more freedom from limitations and regulations. Against official rules and some cultural norms, representations of women in social media businesses resemble those from western cultures. Moreover, some women business owners believe social media might be more accessible for their informal commercial activities. These interviews highlighted the desire of many people who solely operate on social media to establish a physical location in a city if certain conditions (e.g. having enough money to start a business, or some regulations being changed) are met.

Even though for some people social media is a complete alternative to public space, a metaphorical experience of commercial activities in urban space is evident. Many businesses use names of specific bazaars such as Tahlenji or the Friday Bazaar to quickly communicate certain products or qualities. Some businesses refer to themselves as street vendors without having a physical location to potentially communicate inexpensive prices to customers. Pages that sell women's products refer to the regulations of urban spaces through digital 'Men Are Not Allowed' signs. Even audible experiences of second-hand electronics vendors can be seen on various social media pages.

Other analogies to the dynamics of informal commercial activities in urban spaces can be found. Many cities in Iran and around the world show stark class differences between where neighborhoods of the wealthy, high-end businesses are located and where poor populations live. In social media, capital could be manifested through the number of followers. Celebrities, influencers, famous politicians, and large established businesses enjoy millions of daily traffic and make money through advertisements, while a majority of pages have a smaller number of followers. Analogous to how street vendors go to high-traffic wealthy areas of cities, small online vendors advertise their pages under comment sections of big accounts to take advantage of their high volume of visitors. Although social media platforms offer new ways for poor populations to make a living, the platforms allow popular pages to become more prosperous, leaving traces of the economic inequalities in cities on the social media space.

Although social media offers new freedom to some disenfranchised populations, many populations have less access to these benefits and freedoms. People in extreme poverty or in rural areas without proper access to computers, smartphones, and the internet are completely disconnected from its benefits. Moreover, individuals with disabilities, who might not have easy access to the streets of the city, might face issues with interfaces that are not designed to be accessible to everyone. Finally, in multi-ethnic and multilingual communities such as Iran, minorities who speak other languages might have less of a presence and a harder time engaging in business.

Unfortunately, women and the LGBTQ population are harassed and troubled as they walk about and engage in life within their cities. This problem is arguably more present on social media. Due to a smaller regulatory presence (especially for Non-English-speaking communities) and the relative anonymity of identities, women and other disenfranchised people face often harsher harassment on social media spaces.

This chapter highlighted how social media can be seen as a relatively new, important, and impactful element within the context of informal economies. Social media provides alternatives to urban spaces and has many of the problems that exist in urban spaces. The poor populations intelligently use social media in conjunction with the streets of the city to make slow but steady political and economic digital progress, or in similar terms of Bayat: "a quiet **digital** encroachment of the ordinary".

Notes

1 https://srtc.ac.ir/analytical-reports/جمعیت-به-نگاهی-با-آن-آینده-و-ایران-شهری-جمعیت کلانشهرها.

2 Bayat, Asef. (2000). "From dangerous classes' to quiet rebels' politics of the Urban Subaltern in the Global South." *International Sociology, 15*(3), 533–557.

3 Bayat, Asef. (2012). "Politics in the city-inside-out." *City & Society, 24*(2), 110–128.

4 Bayat, Asef & Kees Biekart. (2009). "Cities of extremes." *Development and Change, 40*(5), 815–825.

5 Bayat, Asef. (2020). *Life as Politics.* Redwood City: Stanford University Press.

6 Bayat, Asef. (1997). *Street Politics: Poor People's Movements in Iran.* New York: Columbia University Press.

7 https://www.tehrantimes.com/news/431713/Some-64-of-Iranians-are-internet-users-report, https://datareportal.com/reports/digital-2020-iran.

8 https://www.businessinsider.com/iranian-apps-that-mimic-western-counterparts-netflix-uber-app-store-2019-10.

9 https://web.archive.org/web/20200926025741/https://www.en.eghtesadonline.com/Section-technology-13/29059-iran-most-popular-android-apps.
10 https://web.archive.org/web/20210203134829/https://www.tasnimnews.com/fa/news/1399/09/16/2404577/
11 https://www.nytimes.com/2018/05/01/world/middleeast/iran-telegram-app-russia.html.
12 https://www.bbc.com/persian/iran-55827159.
13 https://www.bbc.com/persian/iran-55827159.

8

UNDERGROUND RESTAURANTS

Pop-up and underground restaurants also referred to as supper clubs are social dining events that operate out of private residences or other spaces not habitually used as restaurants, often ignoring or breaking zoning and health regulations. These eateries are organized by established chefs and culinary entrepreneurs and may be attended by eight to fifty patrons, unknown to each other.

Restaurants operating outside of traditional brick-and-mortar establishments have a long history, including, for example, "Casa Particulares" on the isle of Cuba, which inspired the evolution of contemporary underground restaurants. Operating out of one's home, guests are provided overnight accommodations and meals and receive recommendations to stay in other casas. In the United Kingdom, a couple was inspired to create Brovey Lair in their Norfolk home which is rated as the best fish restaurant in the region.[1] The restaurant is a room at the back of their house with four extendable tables; there's a view to their garden and the stove, fridge, and worktop. Underground restaurants are located across the world, including South Africa, Holland, Russia, San Francisco, and Chicago.

The popularity of underground restaurants varies with location and over time. Young chefs welcome their lower barriers of entry and temporary operations over the considerable capital and long-term investment required for a traditional restaurant. Following the 2008 US recession, culinary entrepreneurs increasingly turned to part-time work and start-up organizing that allowed for self-employment and flexibility. For experienced chefs, these restaurants are a chance to freely experiment with new kinds of food and new audiences. For diners and chefs, the location of the underground restaurant can remain fixed or move to multiple locations (castles, wine cellars, art galleries)[2] For diners, it provides the opportunity to try emerging cuisines or have a unique social experience with strangers.

DOI: 10.4324/9781003026068-10

Underground restaurants are typically shared by word of mouth, guerilla advertising, and the use of various forms of communication technology
ranging from an email notification to social media to websites such as Bon-
Appetour and Eventbrite. Chefs prefer not to share their location publicly
on social media due to limited seating capacity and the potential for conflicts with local regulations that prohibit their business activity in private
residences. Instead, they rely on email notifications to notify patrons that
have confirmed reservations. However, images of meals and the patron's experiences are used for public distribution on social media. Images provide
a variety of information about the quality and type of cuisine, the spatial
qualities of the location, and the social activity taking place during the supper. Images are typically not geolocated to maintain location anonymity.

The success of underground restaurants fundamentally depends on the
chef's ability to offer an immersive dining experience to their audience. Sociologist Daphne Demetry explains that the experience is dependent on the ability of the chef and diners to co-produce a sense of authenticity. Authenticity
is defined as patrons' subjective perceptions of the chef's external expressions
as genuinely representing identity.[3] Authenticity occurs when audience members' experiences with a product or producer align with the organization's
identity claims. In these dining experiences, Daphne argues authenticity
occurs with three illusions: community, transparency, and gift-giving. The
illusion of community is characterized by an ephemeral state of camaraderie
and togetherness, whereby diners who are strangers refer to one another as
friends or even family. The chefs use rhetoric, communal tables, homey aesthetics, and selective membership lists to promote community. The illusion
of transparency is produced by selectively revealing aspects of the cooking
process and minimizing the visibility of elements of the process that may be
incompatible with the illusionary experience associated with an "all-access
pass" that encourages patrons to ask questions or slip into the kitchen to
view the production process. The illusion of gift-giving occurs when the chef
downplays and distracts audience members from the economic transaction
that grants them access and without which there would be no underground
restaurant; instead, they provide the illusion of gift-giving by using alternative payment devices such as envelopes for cash.

Working from Daphne's research on the social construction of dining
experiences surrounding community, transparency, and gift-giving, this
research investigated the specific role of space through imageic imagery
posted on social media. Underground restaurant spaces are unique settings
within a private residential home or backyard, local art gallery, or empty
warehouse space. Within the space, a chef often hosts his guests at a single
large, communal table that encourages social interaction and communal
eating. Additionally, chefs enhance the spaces with lighting, furniture, and
dishware that play a role in a patron's experience. Images offer clues about
the distinctive spaces and become important in constructing a patrons' experience and interest in the event. In some cases, a patron's first encounter
with a dining space occurs through images. In other words, patrons can

develop an imagined reality of the space using imageic imagery well before the event takes place. For this reason, chefs are aware that the process of imageing and publicizing the dining space on social media is a critical part of the patron's future experience.

Data Analysis

The analysis of dining spaces included three approaches to analyze space in social media content. The first approach involves a cluster analysis of chef and patron images on social media that generates clusters of images using the similarity of identifiable features (i.e. shapes and objects) determined by the computer algorithm. The second approach involves a hashtag network analysis that investigates the co-occurrence of hashtags found in the posts on social media. The third approach involves a qualitative thematic analysis of chef's images of their dining spaces using text-based messages to arrive at a set of topical themes.

The cluster analysis began with collecting 7,089 images posted by chefs and patrons in July 2020. Using K-means clustering, a method to partition observations (In this case, ImageNet image embeddings) into clusters based on proximity, five clusters of images were identified (Illustration 8.1). The clusters varied in the homogeneity of the images, but after visual inspection was labeled as spatial setting, ingredients, cooking, plating, and plate as aesthetic objects.

ILLUSTRATION 8.1 Cluster analysis of #Undergroundrestaurant images.

The images in the spatial setting cluster were more mixed than other clusters but had several clear themes. For example, a number of images included menus or other printed material, whereas others showed wine bottles or beverages. But the most common images among this cluster were of the table showing the dinners grouped around a common table. These were shown not just in terms of the arrangement but also in terms of lighting focusing attention on the table within a relatively darker space. The images in the ingredients cluster showed various raw foods prior to being cooked or in large pans before plating. The images in this cluster emphasize large quantities of food. The cooking cluster images focused more narrowly on the cooking process, often on the application of heat to individual portions of protein, grinding spices, or stirring a pan of sauce. The plating cluster showed the process of plating the food focusing not on a single plate but the process of adding ingredients often to several plates. Both the cooking vessels and the plates were present in the images. Finally, the plate itself as an aesthetic object was remarkably uniform in perspective and focus. Almost every image was of a plate presented from a top-down view with the circular form of the plate filling the frame perfectly.

This analysis shows there are common ways chefs and patrons capture images to illustrate the underground restaurant experience. Food is central to the experience. The multiple groups of images relative to the ways chefs choose ingredients, prepare and cook the meal, and present the meal reveal the importance of food in underground restaurants. More unique is the prevalence of communal dining in images which suggests social engagement is significant to the experience. A close look at this group of images shows the type of space (living room, wine cellar) and the ambiance (low lighting, rustic furniture, bright floral arrangements), and types of social engagement (eating, talking, and posing for a image).

Hashtag Network Analysis

Using the data set in the first approach (i.e. 7,089 images posted by chefs and patrons in July 2020 using the hashtag #undergroundrestaurant), the second approach involved a hashtag network analysis that detects the frequency of more than 10,000 hashtags. In this model, hashtags are nodes and the co-occurrences of these hashtags within Instagram posts are links between them. The result is a web of clustered hashtags organized by how frequently they are posted with other hashtags, also known as their co-occurrence. Using Gephi to visualize and conduct the Louvain community detection algorithm,[4] several communities of hashtag usages were extracted. Illustrations 8.2 and 8.3 show hashtags with close proximity are more frequently posted with other hashtags. The line segments connecting the hashtags indicate whether a pair or group of hashtags occurs in a single post.

After constructing the network and community detection, the quantitative analysis found 122 clusters of hashtags. The clusters have a wide range of total hashtags ranging from 1 to 1,350, however, only 18 clusters included 100 or more hashtags and were prioritized for further analysis. These 18 clusters included hashtags ranging from 1,350 to 172 hashtags in a cluster.

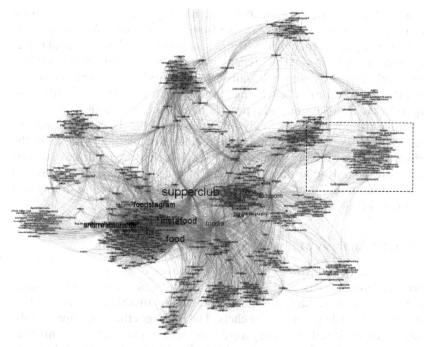

ILLUSTRATION 8.2 Network analysis of underground restaurants social media.

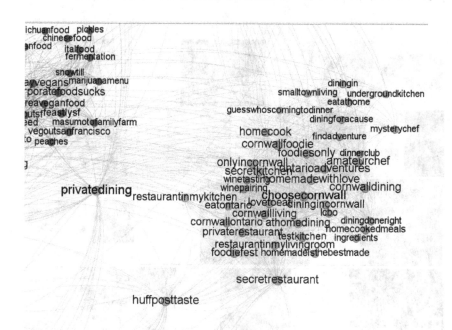

ILLUSTRATION 8.3 Detail of Network analysis of underground restaurants.

Next, each cluster was examined for the hashtag content. Clusters show a diversity of descriptive topics ranging from types of chefs (Michelin, home cook, private), different cuisines (Chinese, Japanese, Italian), types of ingredients (organic, farm-to-table, vegan), urban locations and countries (D.C., New York City, Los Angeles, Dallas, Lisbon, Italy), and types of spaces (private home, home kitchen, living room). Two clusters emerged with hashtags referencing locations and types of spaces more frequently.

Unlike the image cluster analysis that revealed the processes of making the food such as the preparation of ingredients and plating of the food, the network analysis reveals details about the locations and cultural cuisines that are offered. Rather than expressing how the food is made, the hashtag topics reveal where a patron may expect to find underground dining and what type of culinary foods they may enjoy.

Thematic Analysis of Spatial Settings in Social Media

The third analysis involved collecting 15 images of the restaurant dining spaces on Instagram using manual observation after searching with the hashtag #undergroundrestaurant. The posts including these images (Illustration 8.4) were primarily generated by chefs. The messages that accompanied the images were reviewed to identify words and descriptions that may contribute to the construction of a dining experience. Qualitative thematic analysis was performed on the phrases and hashtags found within the messages.

ILLUSTRATION 8.4 Instagram posts focused on dining table

First-order coding of the messages identified 85 distinct words and phrases relating to space, objects, or people. For example, chefs posted messages that relate to spatial qualities such as "space alive again with people eating and drinking", "small space", and "beneath the trees". Words and phrases relating to the spatial location included "set along Brisbane Corso", "East Side London", and "private home". References to social aspects included "guests come as strangers and leave as new friends", "connect over food", and "community". Objects mentioned included "canopy of lights" and "candle-lit tables".

The first-order codes derived from each Instagram post were then further categorized into second-order codes with one of six topics that best describe their content. The topics included setting (20 total mentions), food (12), experience (22), decor (6), table (8), and social (18). This analysis shows that descriptions about the dining setting and experiences within that setting dominate the discourse. The setting topic included mentions directly related to the characteristics of the spaces such as private location, geographic location, unique features in the location (i.e. tree canopy, fire, small size), and the number of people. The experience topic included mentions about the auditory qualities (i.e. glasses clinking, corks popping, laughter), experience adjectives (i.e. magical, holistic, authentic, spontaneous, unique), and social mood (i.e. comfortable, ease, no agenda). The social aspects described were another frequent topic which included the dynamic between people (i.e. earnest conversations, human connection, shared dining, community) and the type of people (i.e. locals, new friends, private members).

Description of the spatial setting, patrons' experience, and social connections found in messages of images depicting space suggest that an understanding of space in the development of the underground restaurant scene on social media is multifaceted and carries multiple meanings. These physical, social, and experiential aspects of space are interrelated and support the construction of patrons' and the chef's anticipated underground dining event.

Social Media Constructing Experiences

This work demonstrates the active role of social media in constructing an experience of underground dining. As opposed to word of mouth, social media plays a role in constructing a mental image. Food, plates, people, settings create a visual and descriptive language about underground restaurants that is consumed by existing and potential patrons.

Similar to Daphne's research on the social construction of dining experiences surrounding themes of community and transparency, patterns of images and messages addressing a communal gathering and behind-the-scenes food preparation could be identified. Daphne's theme of gift-giving as a transaction between chef and patron was less obvious in the data. Different from Daphne, this research identified spatial settings in the psychological construction of an underground restaurant experience.

Spatial settings in this work show a variety of meanings based on the different analytical methods used. The cluster analysis of images revealed the functional and social importance of a communal table. This simple dining furniture fits the needs of small groups well and enables their participation in establishing social relationships and acquaintances. The table plays a role in fostering the social relationships among participants and gives a collective sense of identity in the experience that is unique to underground dining. The hashtag network analysis showed that the location of the dining space is important to patrons. Their experience is part of a place within a city or country and the dining space happens in a private setting of the home. Finally, the thematic analysis of spatial setting posts showed that the features in the setting, sensory experiences, and social relationships are directly linked to descriptions and images of space. In this analysis space carries physical, experiential, and social meaning.

The role of space in this work can be related to Lefebvre's theory of social space. Lefebvre's concept of conceived space realizes how space is depicted and represented through a subjective, often expert lens. In this work, images and messages surrounding underground restaurants are primarily generated by chefs who can curate the image quality and messaging related to their business. Intending to make their enterprise attractive, chefs pay careful attention to the presentation of food, image quality, and descriptive language that describes various features of their business. This enhancement of reality is conceived space generated by the expert chef who seeks to offer a particular experience.

Lefebvre's perceived space focuses on the physical setting of social performance operating as both a medium and outcome of human activity, behavior, and experience. In this work, establishing social connections while dining is the readily apparent behavior found in social media messaging. This is clearest in messages about making new friends or meeting strangers. Finally, Lefebvre's lived space speaks to the struggle of power that unfolds in space. In underground restaurant dining, the primacy of the private home or dining space to host events underscores the covert reality of this practice that seeks to avoid the formal structures of the restaurant industry (i.e. brick-and-mortar buildings, a large number of patrons, health code inspections, high overhead costs). To this end, space plays a critical role in the construction of underground dining experiences and practices. The development of multifaceted experiences on social media draws similarity to other emergent forms of dining such as food trucks or pop-up farmers markets.

Notes

1 Lanchester, John. "Restaurant: The Cafe at Brovey Lair, Ovington, Norfolk". *The Guardian*. Published March 30, 2012. Retrieved 30 December 2015. https://www.theguardian.com/lifeandstyle/2012/mar/30/cafe-brovey-lair-restaurant-review

2 "A SecretEats Dinner at Spice Route". *Crush Mag Online*. 2015-08-24. Retrieved 15 May 2016. https://crushmag-online.com/a-secreteats-dinner-at-spice-route/

3 Demetry, Daphne. (2019). "How organizations claim authenticity: The coproduction of illusions in underground restaurants." *Organization Science, 30*(5), 937–960.
4 Blondel, Vincent D., Jean-Loup Guillaume, Renaud Lambiotte, & Etienne Lefebvre. (2008). "Fast unfolding of communities in large networks." *Journal of Statistical Mechanics: Theory and Experiment, 10*, P10008.

PART 3

Art and Culture

9

BANKSY

Banksy is an anonymous street artist, active since the early 1990s.[1] His work is often subversive and political, embracing unauthorized work in public locations that questions the role of art and protest. He does not sell photographs or reproductions of his work, although his work is sometimes removed and sold by the property owners or by early arrivals on site. His early work had no advance notice publicly but today news of his work spreads quickly through social media.

Girl with the Pierced Eardrum appeared in a back alley in Bristol, England on October 21, 2014. News of the event spread rapidly, primarily through images posted on social media. It was a large-scale stenciled image based on *The Girl with the Pearl Earring* by Vermeer.

Better Out Than In was Banksy's one-month residency in New York City in October of 2013. He chose New York City for its high foot traffic and many hiding places.[2] During this residency he introduced one piece of art each day (31 in total) at sites in all the five boroughs of the city. These works range from stencil graffiti familiar from his previous work, installations that occupied vacant lots, mobile and vehicular-based art, and a one-day anonymous art sale of previous Banksy prints. This project marks the expansion of the range of his work to include multimedia pieces and letters to the editor. Unlike earlier work, Banksy's dedicated website and Instagram account documented and announced clues to locate his art in this project.[3]

Considerable controversy accompanied Banksy's work in New York. While there were eager supporters who arrived promptly at each new site, the mayor was openly hostile to what he termed Banksy's "vandalism". Works featuring flower-strewn images of the World Trade Center attack based on his critique of the banality of the new World Trade Tower and his depiction of a child spray painting in the South Bronx were particularly controversial.

DOI: 10.4324/9781003026068-12

The connection between Banksy's work and public space is in some ways straightforward. Each installation is in a particular place chosen by the artist, based on a specific setting but usually one without prominence (e.g. *Night Vision Horses* on a vacant lot on Ludlow Street, the mural on a back alley in Port Talbot). Banksy's guerilla tactics are in keeping with his identity as a graffiti artist, operating without approval or forewarning and with controversial standing in the world of fine arts.

Given Banksy's desire for provocation as a method to uncover the political nature of art, he has been very successful in gaining wide coverage in the popular press.[4] The focus of this reporting has been partly the description of the art and partly the reaction from a wide variety of residents from art fans, to local residents, to the mayor.

Ulrich Blanché in *Banksy: Urban Art in a Material World*[5] discusses Banksy's work as a form of consumption. After an explanation of the historical context of London at the end of the 20th century, he analyzes the background of consumption as a philosophical and political position citing both Karl Marx and Madonna. He understands consumption not as a solely pejorative term but also as a form of interaction between the artists and the setting, "...consumption has the same relationship to capitalism as praying to religion". He then discusses work by Banksy and the use of covert (and usually illegal) installations, direct engagement with the public and social media, and public discussion as methods of art consumption. He notes, in particular, the requirement of street art to engage with the urban context of its installations.

The purposely obscure and quasi-illegal nature of the installations in urban space leaves unclear issues about ownership and protection of each work. Almost every installation generates complicated legal issues. Who can sell what are immediately very valuable works? Who can move or protect them? Who decides who receives the funds from their sale? These are issues not separate from the creation of the art but rather a part of the performance. After noting the "factual difficulties" surrounding street art, an article in the Chicago Law Review[6] discusses possible legal doctrines to resolve issues of ownership and fragmentation, settling on equitable division as the best and most flexible approach. Equitable division is recommended because of its flexibility in handling difficult cases with rigid standards. One imagines that Banksy would be pleased by the difficulties he is causing the legal system.

The mystery of Banksy's identity has been the focus of research[7] that uses a mathematical technique (Dirichlet process mixture modeling) from the field of criminology to study the pattern of artworks in Bristol and London. Combining the dates and locations of the art with dates and locations of an artist suspected of being Banksy, this article finds probable cause to believe their suspect is the artist (or is it offender?). Given the generally transgressive quality of Banksy's work, a cat-and-mouse game with the police might be part of the game plan.

Hansen and Danny[8] discuss Banksy's work *Slave Labour* in North London which was extracted from a wall, moved to Miami, and listed for auction. The recognition of Banksy as the artist led to this extraordinary effort and to the expectation of significant financial gain (the work did eventually sell for £750,000). Beyond the irony that the work referred to the type of child labor that produced the souvenir items on sale adjacent to the site as part of the London Olympic games, it set off a sequence of installations by local street artists on the site, followed by efforts at valuation by local authorities. Installations occurred five times and were followed, in rare cases, by preservation or, more commonly, erasure. Using Ranciere's[9] concept of division of the sensible from the insensible, this can be understood as a political struggle to control what is art and worthy of aesthetic appreciation. Insurgent art struggles for recognition with the established order. Aesthetics is central to this battle because it takes place over the image of society – what is worthy of being discussed or portrayed.

Turning to the issue of the role of urban space in Banksy's art, Hochman, Manovich, and Yazdani[10] have researched Banksy's *Better Out Than In* New York residency using the conceptual framework of hyperlocality. Hyperlocality arises from the precise location in space and time of social media data used, for example, for microtargeting in the fields of advertising and media. This approach allows the study of narrowly defined locations and time periods, for example, events on a single block that last only a day or an hour. There is an obvious connection to the 29 installations of *Better Out Than In* where the audience is led on a month-long scavenger hunt across the city of New York. This research proceeds to collect 28,000 photographs with hashtags relating to Banksy from social media. Clustering algorithms are used to sort the photographs and display them in x, y matrix plots using combinations of two factors at a time, for example, location and time, brightness mean and hue mean, etc. The result is a large display that shows variation relative to the official posted photographs and explores the idea that Banksy's work is unique in combining spatial and temporal aspects as vital to the audience's experience.

This chapter is focused on the influence that social media has on the perception and occupation of urban space. This investigation includes the integration of social media as part of the installation, the timeline of the social media postings, and an analysis of the corpus of photographs using clustering algorithms to identify how social media influences the perceptual structure of the space.

The Girl with the Pierced Eardrum, Bristol

Data associated with *The Girl with the Pierced Eardrum* in Bristol was collected for the month following its installation on Oct 21, 2014.[11] A total of 500 images were collected from Instagram for the following month including home geolocation of the person posting, content of the photo, and time of the posting.

Illustration 9.1 shows a spike in Instagram postings that remains high for the first month and then tapers off by the end of the month. Based on this data, it is possible to track the geography of the spread of the user's home location over time. Illustration 9.2 shows the location of users over a three-day interval who posted images on Instagram after installation. The spread from locations tightly grouped in Bristol to a large number of postings located in London reflects in part the pace of the news spreading over social media. It also reflects the role of London as the center of fine arts in England which drives the interest in Banksy's work. Because this work was not announced but was slowly discovered and shared through social media, it is possible to understand the spatial and temporal features of social media as a collective activity.

The role of social media for the *Season's Greetings* in Port Talbot is also that of a crowd-sourced process of discovery. Banksy did not announce the location of the installation. It was discovered serendipitously and word spread quickly through Twitter and Instagram. Within a few hours it was trending on both platforms and led to both online comments and a large volume of in-person visits to the site.

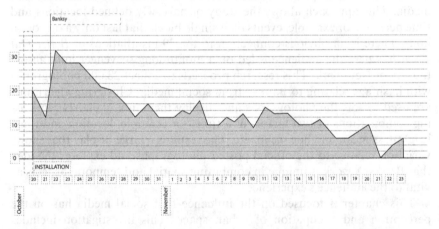

ILLUSTRATION 9.1 Bristol Instagram postings month of Banksy installation.

Day 1 - Bristol   Day 2 - Bristol   Day 3 - London - Bristol

ILLUSTRATION 9.2 Locations in Bristol of Banksy Instagram postings on first three days.

Better Out Than In, New York City

Banksy's decision to be an artist on the street rather than in the museum does focus attention on the importance of understanding urban space. The setting (or perhaps it is even the medium) of Banksy's work is intimately bound up in urban space and uses social media (Instagram) as a daily means of announcing and directing viewers to each installation. In the case of *Better Out Than In*, the 31 days of residency can be at least partially understood through a typology of urban sites. Seventeen of the 31 installations (days 1, 2, 3, 4, 7, 8, 10, 12, 14, 15, 17, 20, 21, 24, 27, 28, 30) are placed in a setting familiar to all street art. An example on day 20 is *Hammer Boy* on the Upper West Side depicting the silhouette of a child hitting the fire warning with a mallet much like the ones found at fairs. It is painted on a blank wall away from the main focus in a location familiar in Banksy's previous work and to graffiti artists generally. These spaces are regarded as secondary or tertiary to the spatial definition of the city, sometimes by being set back from the corner and sometimes being placed in an alley. Indeed, Banksy's work at Port Talbot is located in a literal alley filled with garages and workshops and outside proper building fronts.

Another group of six installations (days 9, 18, 21, 22, 25, 31) occupy three-dimensional sites either on storefront spaces or on empty lots usually enclosed by chain-link fences. *Night Vision Horses* on day 9 was painted onto a car and truck in an empty lot on the Lower East Side showing horses wearing night vision goggles during a shootout between armed men. Compared to the silhouette work mentioned above, the engagement with urban space is similar but modified by issues of scale or degree of public access. Other projects deal with urban space differently. Three projects (days 5, 11, 26) are mobile, mounted, or painted on trucks and moving through the city. One example is *Mobile Waterfall* on day 6, an old delivery truck converted into a mobile garden, featuring butterflies, a rainbow, and a waterfall that toured throughout the month. Although the lack of fixed locations creates a different reading of urban space, they all are seen and understood relative to some urban space, and the adjacency to the locations that trucks can occupy. Two other installations (days 13, 16) are considered the best performances, consisting of an actor and portable props that are temporarily installed. An unnamed example on day 13 is a stall set up in Central Park without notice selling original signed Banksy canvases for $60 each. Most of the work went unsold. These installations occupy locations that are recognizable and understood together with the actor and props as a place, an idea focused on the meaning and occupation of space for a particular purpose. Finally, there are three installations (days 6, 19, 23) that are solely media events on YouTube or Instagram. On day 23, after reports that the Police Department was looking to charge Banksy with vandalism, *Today's art has been cancelled due to police activity* was posted on Instagram. The importance of social media in every installation is highlighted in these cases by the absence of the urban space. Seen within the context of the entire month-long series, viewers infer the existence of urban form by its absence in these particular instances.

During the *Better Out Than In* installations in New York City, the role of social media was engaged by the artist as an active part of his work. Each day, a post on the artist's website announced the location of that day's installation, which led to a rush of visitors and postings on Twitter and Instagram. The only exception to this pattern was the spray art booth set up in Central Park on day 13. This unannounced event consisted of a selection of Banksy's work, of which only three were sold during the entire day. The value of the work was established at over $100,000 by the sale of the few that were purchased but was ignored by thousands of pedestrians who passed the booth.

Using LDA topic modelling on a Twitter dataset of posts related to the hashtag #banksyny highlights social media conversations around Banksy's work (Illustration 9.3). Many of the topics refer to Banksy and his art with words such as print, stencil, graffiti, artwork, and urban art (Topics 2, 6, 8, 9, 14, 15, 19, and 20). Other topics refer to the work within the context of New York City with terms such as banksynyc, and Newyork (topics 13 and 16). Other topics refer to specific art pieces and their locations with words such as dismaland, boy, betteroutthanin and neighborhoods such as Chelsea, Bronx, Manhattan, and Queens (Topics 1, 3, 7, 11, and 17). These topics, in general, hint towards the reactionary nature of the Twitter conversation on the Banksyny work which is partially because Banksy might have intended for this type of social media trending to happen.

Topic 1	banksyni, chelsea, nofilt, signofbanksi, banksybabi	Topic 11	brooklyn, tag, truck, back, greenpoint
Topic 2	banksyart, art, print, sign, banksi	Topic 12	banksyart, artist, art, love, boy
Topic 3	bronx, osgemeo, side, west, east	Topic 13	banksynyc, bikeographi, biciografia, parti, bikefriday
Topic 4	banksyart, two, dismaland, rat, book	Topic 14	banksi, art, found, gricbenaimpres, birthday
Topic 5	piec, one, see, just, work	Topic 15	streetart, art, graffiti, newyork, newyorkc
Topic 6	art, artist, paint, artwork, banksyart	Topic 16	new, york, graffiti, banksyart, art
Topic 7	nyc, manhattan, les, reaper, chasingbanksi	Topic 17	banksynyc, banksyart, queen, tribeca, octob
Topic 8	banksyart, banksystreetart, banksystyl, banksyn,	Topic 18	betteroutthanin, day, today, better, wait
Topic 9	urbanart, graffitiart, graffiti, art, stencil	Topic 19	exhibit, banksi, amaz, instal, banksyexhibit
Topic 10	galleri, banski, art, artbasel, museum	Topic 20	mural, freezehradogan, artist, paint, zehra

ILLUSTRATION 9.3 Topic modelling for Banksy's *Better Out than In.*

An interesting aspect of Banksy's use of social media in New York City is the creation of an online community that becomes a in-person local community that emerges when fans make pilgrimages to new sites throughout the month. Banksy's fans are anxious to get to a new site quickly before it was defaced or removed. Fans often met in person each day, trading information and greetings at each new location.

ILLUSTRATION 9.4   Instagram photo clusters *Hammer Boy* and *Heart Balloon*.

The connection between social media and urban space is an essential element of the artist's work. Social media gave importance to specific often peripheral locations in the city, and drew people and commentary to them. As a part of Banksy art, he used social media to direct attention to locations, anticipated social media as a form of publicity generated after discovery of a location, and questioned the authenticity of a work of art without social media providing provenance.

In addition to the tight interconnection of social media and urban space for individual installations, the shifting focus between the widely spaced installations drew attention to the temporary nature of each individual event. This is in keeping with the experience of many events driven by social media ranging from flash mobs to political demonstrations. Virág Molnár's research on Flash mobs[12] investigates "...the intersection and interaction between new communications media and changing uses of physical urban space" as a new form of sociability. While Banksy undoubtedly has intentions beyond just the playfulness of flash mobs, he harnesses this shifting focus as an artistic "palette". *Better Out Than In* is a clear and forceful case for the temporal as a central concern.

Illustration 9.4 is an analysis of photographs using Banksy hashtags and clustering algorithms that helped to identify photographs of each installation and how social media influenced the perceptual structure of urban space. From 28,000 photographs, a total of 50 clusters were identified. These clusters identified specific locations for the viewing and appreciation of an installation. There are considerable consistencies in the angle of the view and image size revealing preferred viewing positions.

Hammer Boy is located on a blank wall adjacent to a fire standpipe. At the top of this illustration are photos from the two clusters identified for this site. One group of photos were taken from close to the work, perhaps 5 feet with a normal lens, and the other group from perhaps 15 feet (the width of the sidewalk is 17.5). This phenomenon of creating shifts in the perceived space is not unique to social media. As Kevin Lynch has noted,[13] the placement of landmarks has the same effect at scales ranging from the Eiffel tower to an equestrian statue. Illustration 9.5 shows that the shifts in perception created by Banksy's use of social media are

ILLUSTRATION 9.5 *Hammer Boy* refocuses urban space.

distinct in secondary or tertiary urban spaces, drawing attention to otherwise ignored locations. And the perceptual shift can be either temporary or permanent depending partly on the fate of the installation or the shifting interest of the audience.

The case for Banksy's work as ephemeral events extends to other dimensions than popping up in unexpected locations or in the shifting focus of attention within urban space. In many cases, his graffiti is erased, removed, or graffitied over by others. Sometimes it is preserved, appreciated, and even helpful to the local community, as in Bristol where *Mobile Lovers* was identified by Banksy as intended for the Boys Club located near where it appeared and was ultimately sold for their benefit. Whatever the outcome, the work has a life of its own after its placement, one that relies on others, actors, and the translation for meaning and importance. His work immediately destabilizes meaning and continues to provoke more translations over time.

Notes

1 Holzwarth, Hans W. (2009). 100 Contemporary Artists A–Z (Taschen's 25th anniversary special ed.). Köln: Taschen. p. 40. ISBN 978-3-8365-1490-3.
2 Hamilton, Keegan. "Village voice exclusive: An interview with Banksy, street art cult hero, international man of mystery". *The Village Voice*. Retrieved 21 October 2013.
3 Ellsworth-Jones, Will. (2013). *Banksy: The Man Behind the Wall*. St. Martin's Press.
4 Among many others for the New York residency: Ryzik, Melena (1 October 2013). "Banksy announces a monthlong show on the streets of New York".

The New York Times. Retrieved 21 October 2013 https://artsbeat.blogs.nytimes.com/2013/10/01/banksy-announces-a-monthlong-show-on-the-streets-of-new-york/; Barron, James. (15 October 2013). "Racing to see Banksy Graffiti, while it can still be seen". *The New York Times.* Retrieved 21 October 2013 https://www.nytimes.com/2013/10/16/nyregion/racing-to-see-banksy-graffiti-while-it-can-still-be-seen.html.

Buckley, Cara. (28 October 2013). "Monthlong chase around New York City for Banksy's street art". *The New York Times.* Retrieved 3 November 2013 https://www.nytimes.com/2013/10/29/nyregion/monthlong-chase-around-new-york-city-for-banksys-street-art.html.

5 Blanché, Ulrich. (2016). *Banksy: Urban Art in a Material World.* Tectum Wissenschaftsverlag.

6 Salib, Peter N. (2015). "The law of Banksy: Who owns street art?" *The University of Chicago Law Review,* 82(4), 2293–2328.

7 Michelle, V. Hauge, Mark D. Stevenson, D. Kim Rossmo, & Steven C. Le Comber. (2016). "Tagging Banksy: Using geographic profiling to investigate a modern art mystery." *Journal of Spatial Science,* 61(1), 185–190.

8 Hansen, Susan & Flynn Danny. (2015). "'This is not a Banksy!': Street art as aesthetic protest." *Continuum,* 29(6), 898–912.

9 Ranciére, Jacques. (2004). *The Politics of Aesthetics.* London: Continuum.

10 Hochman, Nadav, Lev Manovich, & Mehrdad Yazdani. (2014). "On hyper-locality: Performances of place in social media." *Proceedings of 2014 International AAAI Conference on Weblogs and Social Media,* Ann Arbor, MI.

11 Data for this project in Bristol was compiled by Brendon Bryant in an advanced seminar class focusing on Data, Architecture and the City at the School of Architecture, UNC Charlotte.

12 Molnár, Virág. (2014). "Reframing public space through digital mobilization: Flash mobs and contemporary urban youth culture." *Space and Culture,* 17(1), 43–58.

13 Kevin. (1960). *The Image of the City.* Vol. 11. MIT Press.

10

BURNING MAN

Beginning in 1998, Burning Man changed location from Northern California to the desert of northwest Nevada[1] for the annual nine-day event. With the move to the Black Rock Desert, founder Larry Harvey saw Burning Man as a movement to restore community and creative expression in a time of homogenized mass culture and societal anomie.[2] Burning Man became tightly organized around a set of ten guidelines:[3] radical inclusion, gifting, decommodification, radical self-reliance, radical self-expression, communal effort, civic responsibility, leaving no trace, participation, and immediacy.

The event is a carefully planned temporary city approximately 1.5 miles in diameter organized in a concentric grid occupying 240 degrees of an arc focused on a common area named Center Camp. Burning Man exists temporarily in physical form for only ten days in a year. For the remainder of the year, a collaboration of volunteers using a sophisticated set of websites and blogs create an online community that has an effective way to acculturate participants to the values and goals of the organization.

Larry Harvey founded Burning Man in 1986 and remained active in the organization until his death in 2018. It began with the burning of a single statue of a man on the beach in San Francisco. In the beginning, the event was repeated annually and loosely organized by a motley collection of friends and nonconformists.

Burning Man has been described as a "…techno hippie carnival…which later turned into something of a sprawling frat party for the technogentsia".[4] In 2019, 78,000 people participated in the event including a wide range of age groups, but skewed toward younger, white men. Tickets for the festival cost $475.

The festival features large-scale interactive installation art (including "mutant vehicles") often with kinetic, electronic, and fire elements. All participants need to apply to attend and can also apply for art grants (e.g. in

DOI: 10.4324/9781003026068-13

2006, 29 were funded). The culmination of the event is the Burning Man sculpture, a large-scale 40-foot wooden figure that is set on fire.

Several controversies have surfaced as the festival has grown. One involves the influx of tech billionaires, who have brought both lavish spendings and what is perceived as an overemphasis on the business interest of Silicon Valley not in line with the "gift" ethos of Burning Man. Another is the more and more prominent presence of information and communication technology at Black Rock City marked by the expansion of cell phone service. Disagreements about the open license for all photography and video at the festival are another similar issue.

Burning Man has consistently grown larger and larger. Because of the Covid19 pandemic, the 2020 festival at Black Rock City was canceled. Instead, a virtual event was announced titled The Burning Man Multiverse.

Specifically considering Burning Man before Covid19, Katherine Chen[5] studies the need to maintain organizational focus and enthusiasm about the ten guiding principles during a period when the need to introduce bureaucratic practices can drain the meaning and agency of the members. She introduces the term "charismatizing" to describe this need and cites a number of studies that propose storytelling as a method to maintain enthusiasm toward a shared vision. A story includes three characteristics; past actors and events, a historical sequence, and a common struggle. Stories help contextualize rules and routines, allowing new members to assimilate into the organization and claim agency and understand possible action without a narrow set of rules. Burning Man has a particular structure characterized by in-person interaction for only one week followed by 51 weeks of primarily online preparatory work. Using a qualitative method of in-depth interviews and analysis of online postings, Chen uncovers a widespread use of storytelling that promotes agency and counteracts bureaucratic ritualism.

Dawn Aveline explicitly addresses the use of social media[6] at Burning Man. Using 200 online surveys and semi-structured interviews, and demographic data, she found participants used a variety of social media, including prominently Facebook (which hosts official pages), email, and the JackRabbit Speaks website. Participants report using social media to maintain personal connections throughout the year, to plan and coordinate projects, and to incorporate social media into projects as part of a memorable experience. Additionally, they reported concerns about the commercialization of the festival and the ambivalence about the impact of social media on the festival and the wider society. This work provides an understanding of themes important to the Burning Man community.

Black Rock City

Black Rock City as a temporary, physical urban space has always been a central aspect of Burning Man. This ephemeral city has always served as a way to frame the activities of Burning Man, and also to frame Burning Man in time; it has a clearly defined beginning and end. While the art projects, events, and festivities were of interest individually, they gained meaning from the temporary city in which they occurred. The temporary nature of the city itself resists formalizing the activities of Burning Man long-term and the top-down bureaucratic structures that impede upon society. To this end, what might seem to be odd or even illegal in a more everyday context becomes part of a larger festival constructed explicitly as a counterpoint to mass culture. Art and other events in Black Rock City are understood within the construct of the ten principles of Burning Man. Radical inclusion, gifting, decommodification, radical self-reliance, radical self-expression, communal effort, civic responsibility, leaving no trace, participation, and immediacy exist (more or less) in Black Rock City, and it is reasonable to expect different behavior and different art here than in everyday life elsewhere.

Black Rock City is also unique for its seclusion from other human settlements requiring either an eight-hour pilgrimage from San Francisco over mountains and deserts or a three-and-a-half hours drive from the nearest airport in Boise. The private temporary airport at Black Rock City allows only the wealthiest 1% to circumvent this tedious journey.

Black Rock City also demands a particular kind of attention, just as the cultural, symbolic, and functional presence of a museum leads visitors to evaluate artwork differently than they would in everyday life. Despite the aesthetics of the art itself attracting observers, the environment in which the art lives is enacted and forms part of the follower's experience. *Fountain* by Marcel Duchamp, a readymade urinal placed in a museum in 1917, is understood not as plumbing, but as a comment on the position of art in the contemporary culture. The context of a museum engaged by the Dada art movement provokes a more philosophical position that questions aesthetic, social, and cultural context.

But as important as the framing provided by Black Rock City is to contextualize our understanding of the events and the art of Burning Man, it only partly explains the importance of urban space. The city provides not only isolation but is also a complex and extensive urban network. The experience of a city as large and heterogeneous as Black Rock City is inevitably episodic. This is similar to Guy Debord's goal for psychogeography as a means of renewing our ability to experience the city directly and without preconception. He imagines *flaneurs* walking turtles through the situationists city, accumulating and juxtaposing serendipitous events and experiences.[7]

The layout of Black Rock City began with a reference to gathering around the campfire,[8] with the burning man sculpture (set on fire at the conclusion of each festival) at the center of a circular street grid intersected by radial streets (Illustration 10.1). The rigid geometry reminiscent of Brasilia or a military encampment was largely dictated by access requirements from the Bureau of Land Management for issuance of a permit.

The experience of Burning Man belies what could be an oppressive regularity of the grid. The distinction between the rational organization of the streets is in high contrast to the rich variety of installations, sculptures, and art projects placed in the city. These installations display little of the regularity evident on most urban streets, often dictated by the legal or quasi-legal restrictions on building placement. Most of the camps and installations in the city are free-standing objects that do little to define the edges of the streets. The emphasis on radical self-expression guarantees a wide variety of sizes, locations, materials, styles, forms, and activities. Moving along the 13 curvilinear streets is experienced as an unfolding panorama of energetic or even riotous heterogeneity. The only clear visual organization that is apparent at ground level is the radial street focused on the Burning Man statue at the center.

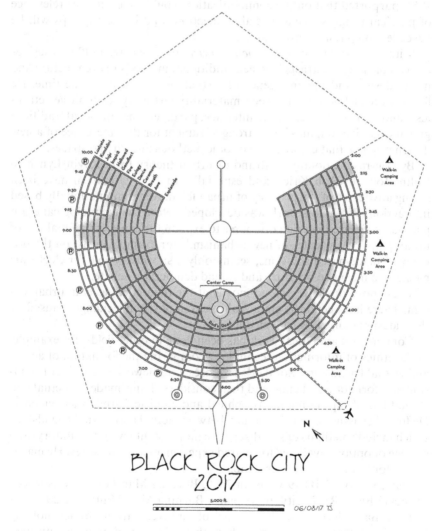

ILLUSTRATION 10.1 Black Rock City plan.

Virtual Burning Man

The shift of Burning Man from a physical location in Black Rock City to a virtual location in 2021 can be situated among scholarly discourse in a number of technological and related fields. The construction of online social communities, non-place urban realms, virtual environments, and cyberspace mark a critical arena of scholarship that seeks to describe the rapid transformation from in-person activities to online communication in everyday life. The global pandemic tested the strength and durability of virtual communication and its infrastructure making it a way of life by rendering it essential today.

Melvin Webber in his iconic piece *The non-place urban realm*[9] written in 1964, purported that online communication would supersede the relevance of physical urban space as social connections and interest groups will be based less on physical propinquity.

Mitchell's *Antitectonics: the poetics of virtuality*[10] written in 1998 describes a series of sharp contrasts between traditional methods in architecture and urban design and the emergence of virtual environments at the time. He focuses on dichotomies between materiality/virtuality, tectonics/electronics, hand tools/software, facade/interface, parti/genome, and local tradition/global organization, making a strong argument for the emergence of a new design practice that embraces electronic methods and virtual spaces.

By contrast, a dominant strand of urban theory from Kevin Lynch to Colin Rowe to Rob Krier, and especially, to New Urbanism have been strong advocates of the primacy of urban form over computationally based interaction. While this work was developed before the technological era it maintains a place in design discourse to explain the experiential qualities of urban places. Advocates of new urbanism,[11] for example, insist on the primacy of place-based solutions, seeing only a subsidiary role for electronic media as a means to organize and extend democratic debate.

These positions have existed mostly at the theoretical level in urban design. There has been little empirical research of specific changes caused by the widespread use of social media.

Comparative analytical work has been done in other fields, for example, in the ethics of information technology comparing the formation of actual and virtual friendships. Søraker[12] proposes a framework to consider friendship as a form of well-being, and then considers claims made for actual and virtual friendships citing philosophical and empirical studies as evidence. He finds key issues that differentiate between actual and virtual friendships which include methods of self-disclosure (non-voluntary and voluntary) and a range of options available for shared experiences and pleasures. He makes claims for both forms of friendship.

Due to the COVID-19 virus, in 2020, Burning Man Festival was forced to cancel Black Rock City. In its place, Burning Man Multiverse 2020 was created. Based loosely on the concept of multiverse from physics (notably Erwin Schrödinger), it imagines a hypothetical group of multiple universes comprising everything that exists: space, time, matter, energy, information, and physical laws. Different universes within the multiverse are called "parallel universes" and are imagined to be simultaneously true. While there is considerable controversy if this is actually a scientific hypothesis, the possible connection to participation, radical inclusion, and radical self-expression appealed to the Burning Man community.

Illustration 10.2 shows five of the eight online Black Rock Multiverse "Recognized Universes" independently created by volunteer members of the Burning Man community. The Burning Man Project provides support, encouragement, and connection through its Facebook site with its 1,370,211 followers.[13]

BRCvr included a complete three-dimensional model of the urban design of Black Rock City configured exactly as it was intended for the site. A complete set of models for the entire city are included. Users were free to navigate throughout the entire city, zooming in and out of details and events including art installations, structures, theme camps, talks, and performances. It was optimized for virtual reality headsets and desktop use.

Multiverse is a full-scale, virtual interactive Black Rock City that includes the deep playa. Members of the community designed virtual communities and sound stages that broadcast live events, workshops, and live music. Artists are translating art installations into 3D forms. Visitors will appear as 3D avatars who can communicate via live voice with others via a mobile phone or VR headset.

SparkleVerse is a 2D web-based Universe that combines features of Google with video pop-ups. One can navigate the map marked with experiences the community has created including performances, theme camps, art pieces, or a trash fence party. Visitors can click on video avatars and ambient audio to join other visitors. It is accessible via any mobile or desktop, uniquely located video avatars and ambient audio.

MysticVerse is an interactive 3D Universe using the site and plan of Black Rock City rendered in a futuristic style. It invites the submission of models

ILLUSTRATION 10.2 Five universes within the burning man multiverse.

for inclusion in the MysticVerse, as well as proposals for events, talks, and art projects. It can be accessed by mobile, desktop, VR, and any operating system.

Build-A-Burn is an interactive digital space that features a comic book style rendering of camps, artworks, and events created by the entire community of visitors within the street layout of Black Rock City. Using only a browser and webcam, visitors will be able to wander in an art-filled playa and interact with others.

Ethereal Empyrean Experience is an eight-sided star-shaped temple that is on Black Rock Playa. It provides for interactive navigation through the temple and allows visitors to leave carefully curated offerings of artworks.

The Infinite Playa is the most photorealistic rendering of the 3D world of Black Rock City. It begins with a model of the Black Rock Playa and invites submission from artists, camps, and inspired individuals to place anything on the site including art rendered with full 3D detail, rides on mutant vehicles, and live performances and talks. It is available via a browser on a desktop computer or mobile device.

BURN2 is a Second Life version of Burning Man. It has recreated Black Rock City in Second Life and encourages the placement of art, structures, and events in that world including past Man builds and authorized 3D replicas. All interaction is through avatars, which makes the interaction between avatars possible. Second Life includes the construction of the world through user participation and simple bold graphics. BURN2 is accessible via any device with a web browser and a Second Life viewer.

These "universes" have a number of elements common to advanced online culture including interactivity, interaction with other visitors, and immersive visualization. All of them use the location of Back Rock City, most of them quite literally, as the underlying structure of their online universe. Some of this is motivated by the long history and clear structure of Black Rock City and its salient absence during the current crisis. The popularity of gaming and virtual worlds have created a common language of moving through 3D urban space as a shared digital experience. Many online virtual worlds such as Second Life and The Sims also have readymade tools for the construction of these worlds, several of which were used in the Black Rock Multiverses.

Another motivation is suggested by the "method of loci", also known as the memory palace. This is a mnemonic strategy dating to ancient Greece and Rome[14] using mental visualizations of familiar spatial environments to recall information (e.g. the sequence of a speech). In the contemporary human-computer interface the use of urban space (known as information space metaphor)[15] is a powerful new paradigm for organizing and connecting complex and heterogeneous ideas and events. This suggests an important role for urban space both in the physical and virtual worlds.

The cancellation of Burning Man at Black Rock City in 2020 led to numerous impromptu gatherings at locations across the country. The culmination of Burning Man has always been the fire that consumes the wooden

Burning Man

ILLUSTRATION 10.3 Burning man local burns 2020.

sculpture at the center of Black Rock City. In 2020 this event was repeated in miniature at these impromptu sites. Illustration 10.3 shows examples of these "mini" burns collected from 1,100 Instagram posts in 2020. This recreates the spatial typology of the festival not as material form but rather as a choreographed set of archetypal spatial acts and locations. In religious studies, Mircea Eliade[16] has described these reenactments of the cosmology and location of the center as necessary to create a sacred space and to renew the world. The examples of the burn as small events in the back corner of a lot or a Weber grill represent a recommitment to the ideals of Burning Man in spatial form.

Burning Man is an example where the urban space of Black Rock City has become an important or even central method of demonstrating its guiding principles. This includes its role as a counterpoint to the vigorous online organization for the majority of the year, its role as an isolated event far from "everyday" cities, its role as a suitably flexible spatial matrix for the heterogeneous content, and its role as a metaphorical structure capable of reenactment in other locations. The importance of this urban form is reflected in the persistence of its use as a metaphor and organizing structure during the COVID virtual online festival in 2020.

Notes

1 "The Extraordinary History of Burning Man". Retrieved 18 June 2020. www.farandwide.com
2 Witt, Emily. "How Larry Harvey, the Founder of Burning Man, Taught America to Experiment." *New Yorker*, May 6, 2018.
3 "The 10 Principles of Burning Man". *Burning Man*. Retrieved 18 June 2020. https://burningman.org/culture/philosophical-center/10-principles/.
4 Holson, Laura M. "How burning man has evolved over three decades." *New York Times*. Retrieved 30 August 2018. https://www.nytimes.com/2018/08/30/style/burning-man-sex-tech.html.
5 Chen, Katherine K. (2012). "Charismatizing the routine: Storytelling for meaning and agency in the burning man organization." *Qualitative Sociology, 35*(3), 311–334.
6 Aveline, Dawn Ellen. (2012). "Mirror and Shadow: Social Media in the Burning Man Community." Order No. 1530848, University of California, Los Angeles. http://rwulib.idm.oclc.org/login?url=https://www.proquest.com/dissertations-theses/mirror-shadow-social-media-burning-man-community/docview/1267138222/se-2?accountid=25133
7 Debord, Guy. (May 1958). "Theory of the dérive." *Internationale situationniste, 2*(20.05), 2015.
8 Rohrmeier, Kerry & Scott Bassett. (2015). "Planning burning man: The black rock city mirage." *The California Geographer*, 54, 23–46.

9 Webber, Melvin. (1964). "The Urban Place and the Nonplace Urban Realm". In M. Webber (Ed.), *Explorations into Urban Structure* (pp. 79–153). Philadelphia, PA: University of Pennsylvania Press.

10 Mitchell, William J. (1998). "Antitectonics: The poetics of virtuality." In J. Beckmann (ed.), *The Virtual Dimension: Architecture, Representation and Crash Culture*. Princeton Architectural Press, New York, 205–217. Although he later makes an argument for the role of place in a virtual world, this article has none of that nuance.

11 David, Walters. (2011). "Smart cities, smart places, smart democracy: Form Based codes, electronic governance and the role of place in making smart cities." *Intelligent Buildings International, 3*(3), 198–218.

12 Søraker, Johnny Hartz. (2012). "How shall I compare thee? Comparing the prudential value of actual virtual friendship." *Ethics and Information Technology, 14*(3), 209–219.

13 https://www.oculus.com/blog/burning-man-2020-goes-virtual-the-universes-of-the-burning-man-multiverse/.

14 Cicero, Marcus Tullius. (1862). *Cicero De oratore*. Leipzig: BG Teubner.

15 Benyon, David. (2001). "The new HCI? Navigation of information space." *Knowledge-Based Systems, 14*(8), 425–430.

16 Eliade, Mircea & Willard Trask. (1959). *The Sacred and the Profane: The Nature of Religion*. New York: Harcourt, Brace & World.

PART 4
Extremism

11

CHRISTCHURCH

On March 15, 2019,[1] Brenton Tarrant, an Australian citizen, entered the Al Noor Mosque at 1:40 pm and the Linwood Islamic Centre at 1:52 pm, both in Christchurch, New Zealand. Using semiautomatic weapons modified for automated fire, he killed a total of 51 people and injured 40 people. Tarrant visited both mosques before the attacks, participating in prayers.

Social media played a key role in Tarrant's radicalization that led to the attacks on the two mosques. He was active on several alt-right social media platforms since 2012.[2] He published a manifesto "Great Replacement" on Twitter and 8chan explaining the rationale for the shooting referring to "white genocide" which is a term linked to the growth of minority populations.[3,4] Tarrant published other tweets with references to white supremacist ideas, including white fertility rates, and crimes conducted by immigrants.[5]

During his attacks, Tarrant livestreamed a video on Facebook Live. The video shows him playing music while driving to the mosque including an anti-Muslim song title "Remove Kebab". Using a camera mounted on his helmet, the video shows him walking into the mosque and opening fire.[6] Despite efforts by alt-right bloggers to disguise the video, 1.5 million copies were identified and removed in the first 24 hours after the shooting by security teams at Facebook.

After the attack, he was arrested and police recovered six rifles and 7,000 rounds of ammunition. The rifles were covered in white writing with the names of people and events of battles, Islamic terrorists, and far-right terrorists. He ultimately pleaded guilty to the murders. He is the first person ever sentenced to life without parole in New Zealand.

Response from the government has taken several forms. Prime Minister Arden attended public vigils at both Mosques, strongly condemning the

DOI: 10.4324/9781003026068-15

shootings. She refused to use the gunman's name in public,[7] instead urging people to use only the names of the victims. She ordered a royal commission of inquiry which issued a report on the attacks. The Livestream video of the attacks was outlawed, and several people were prosecuted. A new law was passed in April 2019 that outlawed semiautomatic and military firearms.[8]

Mosques in general are important community gathering places for Muslims around the world and have played important roles within the Arab-spring protests.[9] For example, out of the 2,600 people identifying as Muslim, the Al-Noor Mosque had 550 members with another 450 who regularly attended mosque events.[10]

Extremist or terrorist attacks often occur in places with symbolic relation to their ideological targets: the twin towers symbolized US ideology, the Al-Noor Mosque was a community space for Muslims, the Pulse nightclub was a place with LGBTQ regulars, as was Kosher supermarkets and Synagogues for Jewish communities.[11,12] In all of these spaces communities had valued them in everyday life as a meaningful part of the community. Both the use and symbolism of space in a community can shift in the wake of an event.

The media coverage after the shooting was diverse. Every-Palmer and colleagues conducted a thematic analysis of news articles in the three months after the Christchurch shooting.[13] Unlike other shooting cases, they found the media articles had very limited coverage of the shooter, whereas a large number of articles focused on the victims, their families, and their community. They also found a positive sentiment towards the government's handling of the shooting and arguments for changes in gun laws. Other articles discussed the ways social media was easily weaponized by the attacker and how platforms failed to regulate the ultra-right-wing rhetoric and video Livestream of the shooter.

Data Analysis

Response to the shooting on social media was massive and global. Immediately after the shooting occurred, there was a spike of social media posts. In order to examine the social media response to the event, more than 690,000 tweets with the hashtag #christchurchshooting, #christchurchshooting, #ChristchurchTerrorAttack, #ChristchurchAttack, #NewZealandTerroristAttack, #NewZealandMosqueAttack, #Christchurchshooter, #nzshooting between March 15, 2019 and April 26, 2020 were collected. The tweets were preprocessed and analyzed using Top2Vec[14] topic modelling. Top2Vec topic modelling can use embeddings from large pre-trained deep learning models to extract topics from the hashtags without the need for choosing a predefined number of topics (See Illustration 11.1).

The most prevalent topic within the dataset was one mentioning either "New Zealand" or "Christchurch". This topic included many tweets with only single or multiple hashtags such as #christchurch that were used to

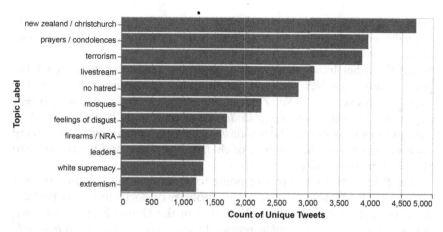

ILLUSTRATION 11.1   Top topics extracted from #christchurchshooting tweets.

trend the news of the shooting on Twitter. Some of the other accompanying hashtags prevalent within this topic included #ChristchurchMosqueShootings and #NewZealandTerroristAttack. Others described the shooting event and other related news.

A large number of tweets described different possible causes for the shooting. Tweets such as *"Terrorism is terrorism....and nothing but terrorism is terrorism....i repeat terrorism is terrorism..."* and *"Shootings...?? What...?? its terrorism, call it terrorism, write it terrorism. #whiteterrorist"* related to the topic of "Terrorism". Other tweets related to the topics "Extremism" and "White supremacy". Tweets reveal the shooter's ideology was close to ultra-right extremist ideologies with statements such as *"The #ChristchurchAttack is a terrorist attack motivated by a right wing ideology"*. Tweets also referenced motivations of the shooter linked to Donald Trump's influence and the use of white supremacists' rhetoric such as *"Trump uses the language of white supremacist whilst condemning the actions of a white supremacist"*.

"Condolences" and "prayers" for the victims and their families were the most important keywords in another prevalent topic. Many tweets such as *"Prayers.Hope. #ChristchurchShooting"* and *"please keep praying. #NewZealandMosqueAttack"* were short and concise. There were also many tweets such as *"May Allah inshallah accept the victims in his heaven"* and *"Today, please make a dua for those people killed at the #NewZealand Masjid"* that expressed condolences and prayers with Islamic vocabulary.

Despite the empathy expressed immediately following the event, others critiqued the tweets suggesting thoughts and prayers were insufficient if no action was taken to prevent shootings in the future. Tweets such as *"It's not enough to offer thoughts and prayers"* were in many cases directed at US politicians and representatives requesting action. For example, tweets directed at the US press secretary or other representatives expressed negative sentiments, *"Yeahhhhhh you work for @realDonaldTrump. Your words mean nothing. Thoughts and prayers don't do anything".* and *"Keep your empty 'thoughts and prayers' to yourself. That is the definition of the least you can do. It's absolutely nothing".*

On the other hand, there was a significant amount of support for the prime minister of New Zealand regarding her strong leadership after the event. Tweets such as *"#NewZealand prime minister @jacindaardern is very impressive! We need more world leaders like her!!!"* and *"Respect to the New Zealand Prime Minister. She is showing true leadership and the right level of compassion"* praised her support of changing gun laws and her response to the Muslim community.

The Prime Minister's rapid response to banning automatic rifles in their country evolved into a debate over gun policies on social media, in particular comments compared the lack of laws in the United States. Firearms, NRA, and gun laws emerged as repeated topics. Tweets such as *"It took only 6 days for New Zealand to ban military style rifles. Over 200 shootings in the US and they're not even close to banning"* and *"This is what happens when your politicians are NOT owned by the #NRA #CommonSense gun laws"* painted a contrast to the United States.

On the other hand, several anti-gun law individuals were using Twitter to argue against bans on automatic rifles. Many of these tweets discussed how such bans will make societies more unsafe: *"The NZ government thinks that banning semi-automatic weapons will solve mass shootings. You're only getting closer to a disarmed and unsafe society. An armed society is a polite society!"*

Mosques or Masjids were prevalent topics of discussion within the Twitter messages. Many of these tweets described the activities taking place in the mosques when the shooting happened. For example: *"Christchurch Mosque shooter attacked a mosque on a Friday, during Jummah prayer, just shows how premeditated this was"*. However, a significant number of tweets were about people who were organizing to go to other mosques in their cities as a symbol of anti-white supremacy and solidarity with the victims such as *"NewZealanders from all backgrounds have started visiting mosques to stand behind and protect Muslims as they pray"*, *"Hundreds pray in the mosque at Lakemba... mourning those who were gunned down for doing the exact same thing"*, and *"This gentlemen in New York walked into a mosque with roses to express his sympathies and condolences for the Muslim"*.

Four of the prevailing topics were mapped using their frequency over a one-year timeline. The graphs show most of the tweets were posted in the immediate aftermath of the shooting. There were two other smaller spikes in activity. The first took place in July and August when the buyback program of guns started to take place. An example tweet around this time is *"#NewZealand begins first round of gun buybacks after #Christchurchattack Are you watching #USA?"*. The second spike in activity occurred around the first anniversary of the shooting when many people remembered the shooting and the victims. For instance, *"A year on from #ChristchurchTerrorAttack during Friday prayers at a mosque and Islamic centre"* (Illustration 11.2).

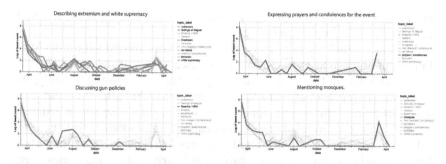

ILLUSTRATION 11.2 Temporal change for four topics related to Christchurch tweets.

Local Community and Meanings of Space

Spurred by the New Zealand mass shooting, Twitter became a public sphere for people from around the world to debate policies.[15] The topic modelling and qualitative evaluation of tweets highlighted discussions that were either an emotional commentary about the event or reactions and retweets to the news coverage. Many of the topics in this analysis aligned with the themes extracted from mass media articles by Every-Palmer[16] and collectively show the mobilizing power of social media.[17,18]

Among this collective effort for change, there were topical focuses among individuals with different cultural or ethnic backgrounds. For example, Muslims who were the main target of the shooter focused on discussions about politicians' supremacist rhetoric about the event. They debated whether such a mass shooting should be called terrorism similar to events carried out by Islamic fundamentalists and extremists. On the other hand, US citizens criticized the Trump administration for their use of anti-white supremacy rhetoric. They also discussed New Zealand as a positive role model for other countries and contrasted it with the lack of change in gun laws and the influence of the NRA in the United States.

Work by Starbird[19] has studied the formation and impact of local online communities in terms of their physical proximity and social media behavior in response to extreme events such as the Boston Marathon Bombings. Using grounded theory, she conducted 11 interviews with participants and found the importance of emotional proximity in structuring online activity and the spread of misinformation on social media. She also analyzed tweets in the aftermath of the Boston Marathon bombing and found that policy debates on social media might be influenced by "emotional proximity" or the nature of how emotionally and culturally close specific communities are to such events.

The place of the event has been shown to be a powerful rhetorical device in guiding social movements.[20] Work in the field of rhetoric by Endres and Senda Cook identified 3 ways that places can behave rhetorically: through a pre-existing meaning of a place, by subverting the meaning of a place, or by permanently altering the meaning of a place.

Mosques, which are important gathering spaces knitted within the Muslim community, played a key role as a place with shifting meaning both locally and within the social media response to the shooting. Locally, the mosques will remain a sacred and celebrated space, but after the event, they will also be recognized as a space of societal significance where religious beliefs and rights are challenged. Radicalized individuals target these local spaces to symbolize their hatred, whereas social media becomes a place for the Muslim community to establish solidarity for a global movement against extremism and supremacy.

Notes

1 "Mosque attacks timeline: 18 minutes from first call to arrest". *RNZ*. 17 April 2019. Retrieved 29 March 2020.
2 Brenton Tarrant: The 'ordinary white man' turned mass murderer". *The Daily Telegraph*. 16 March 2019. Archived from the original on 15 March 2019. Retrieved 16 March 2019.
3 Welby, Peter. (16 March 2019). "Ranting 'manifesto' exposes the mixed-up mind of a terrorist". *Arab News*. Archived from the original on 17 March 2019. Retrieved 17 March 2019.
4 https://www.reuters.com/article/us-newzealand-shootout-internet/new-zealand-mosque-attackers-plan-began-and-ended-online-idUSKCN1QW1MV.
5 https://www.reuters.com/article/us-newzealand-shootout-internet/new-zealand-mosque-attackers-plan-began-and-ended-online-idUSKCN1QW1MV.
6 https://www.nzherald.co.nz/nz/christchurch-mosque-shootings-gunman-livestreamed-17-minutes-of-shooting-terror/BLRK6K4XBTOIS7EQCZW24GFAPM/.
7 Wahlquist, Calla (19 March 2019). "Ardern says she will never speak the name of Christchurch suspect". *The Guardian*. Archived from the original on 19 March 2019. Retrieved 20 March 2019.
8 "'There will be changes' to Gun Laws, New Zealand prime minister says". *The New York Times*. 17 March 2019. Archived from the original on 17 March 2019. Retrieved 18 March 2019.
9 Gerbaudo, Paolo. (2012). *Tweets and the Streets: Social Media and Contemporary Activism*. Pluto Press. PAGE 66.
10 http://www.stuff.co.nz/the-press/10133412/Fighting-killing-not-the-Muslim-way.
11 https://web.archive.org/web/20191210230341/https://www.thejc.com/news/us-news/six-people-confirmed-dead-as-jersey-city-shooting-targets-kosher-supermarket-1.494201.
12 https://www.washingtonpost.com/nation/2018/10/28/victims-expected-be-named-after-killed-deadliest-attack-jews-us-history/.
13 The Christchurch mosque shooting, the media, and subsequent gun control reform in New Zealand: a descriptive analysis, https://www.tandfonline.com/doi/full/10.1080/13218719.2020.1770635?casa_token=eW_F4syflekAAAAA%3AadBJ-mygov-Hh1FZY5C8BQ5D2kBhsVExadKFtv3k_8m1pO4n1hUOokiNsgttSvwRxhrujp0zbOY9.
14 Angelov, Dimo. (2020). Top2Vec: Distributed Representations of Topics. arXiv preprint arXiv:2008.09470.
15 Shirky, Clay. (2011). "The political power of social media: Technology, the public sphere, and political change." *Foreign Affairs*, *90*(1), 28–41.

16 https://www.tandfonline.com/doi/full/10.1080/13218719.2020.1770635?casa_
token=kWirNtrpFAoAAAAA%3AaOf-m12ZD2M6MjDZdrH04NzpPc4lq2
M0ZwxxH85j0Ik_rOu9SBti_IHM-vN2FzEHgZa2Ho_ykMf.

17 Shaw, Donald L. & Shannon E. Martin. (1992). "The function of mass media agenda setting." *Journalism Quarterly, 69*(4), 902–920.

18 Meraz, Sharon. (2009). "Is there an elite hold? Traditional media to social media agenda setting influence in blog networks." *Journal of Computer-mediated Communication, 14*(3), 682–707.

19 Starbird, Kate & Leysia Palen. Pass it on? Retweeting in mass emergency. Proc. of ISCRAM 2010. Huang, Y. Linlin et al. (2015). "Connected through crisis: Emotional proximity and the spread of misinformation online." *Proceedings of the 18th ACM Conference on Computer Supported Cooperative Work & Social Computing.*

20 e.g. Deborah G. Martin & Byron Miller. (2003). "Space and contentious politics." *Mobilization: An International Journal, 8*(2), 143–156.

12

PULSE NIGHTCLUB

Despite their many affordances, social media provides a platform for various kinds of extremist ideologies to flourish. The Islamic State also called ISIS, ISIL, and Daesh,[1] is an extremist Islamist terrorist organization operating primarily in Syria and some parts of Iraq that has influenced terrorist acts in numerous cities and public spaces around the world. Using YouTube, Twitter, chat rooms, and livestreams they are able to easily spread their propaganda.[2] The mass shooting at the Pulse nightclub by Omar Mateen, a US citizen with an indirect link to terrorist organizations,[3] was one of the most horrific terrorist acts within the United States.

Pulse, a nightclub in Orlando, Florida, frequented by the LGBTQ community, was a well-known dance club drawing primarily a Latino audience. On the night of June 12, 2016, it was crowded with around 320 people at closing time.[4] Mateen entered the nightclub with a rifle and a handgun. After failing to stop Mateen, a security guard on duty made a 911 call that eventually brought 100 police officers to the nightclub.[5] In his own phone call to 911 made shortly after he began shooting, Mateen claimed allegiance to the leader of the Islamic State of Iraq, Abu Bakr al-Baghdadi, and said the US killing of Abu Waheeb in Iraq a month before was the reason for his attack.[6]

Mateen opened fire in the nightclub, where dozens were killed or injured in the initial assault. People trapped in the building during the shooting sought help by phone or other means on their smartphones. As he held hostages within the club, Mateen searched Facebook to see if his actions were trending.[7] Social media was also used by police as a mass communication tool during the event. Warnings about false gunfire such as *"That sound was a controlled explosion by law enforcement. Please avoid reporting inaccuracies at this time"* served to calm the fears of the local community members as the police attempted to apprehend the suspect. The hostage standoff lasted for three hours until the police decided to storm the area holding the hostages

DOI: 10.4324/9781003026068-16

by breaking through a wall. Mateen was shot by police officers in a gun battle. Two tweets were posted by police after the event: *"Pulse Shooting: The shooter inside the club is dead"*, and *"Any information you have could aid investigators in this case"*. A total of 49 people were killed and 53 wounded.

In the aftermath of the shooting, there was an explosion of activity on Twitter. To uncover the general topics of discussion on Twitter, a dataset of more than 750,000 tweets with hashtags #orlandoshooting, #pulseshooting, #PulseClub, and #PulseOrlando between June 10, 2016, and September 1, 2016, were collected. The most prevalent topics were extracted using Top2Vec[8] topic modelling and evaluated qualitatively (Illustration 12.1).

The social media sphere of discourse about the Pulse nightclub shooting is full of polarized debates, arguments, and expressions from different communities. Firearms and the NRA are the most salient topics. Some of the discussion revolved around the ability of the shooter to legally purchase guns before the shooting. Some users criticized the gun policies of the United States by pointing out that *"Gunman purchased 2 of the firearms in the last week legally. He bought the handgun and the rifle in last few days"*. Others, defended gun ownership by pointing out the radicalized views of the shooter with tweets such as *"#OrlandoShooting happened NOT because he bought a gun. It happened because he was radicalized. Ban Isis teachings, not guns. #NRA #2A"* and *"Disappointed that Obama used this carnage at such a time to talk about gun control!"*.

Conversations related to the topic of radical Islam included tweets such as "radical Muslim", "radical ideology", #OrlandoShooting", and "Radical Islam hates us all". The hashtag "#wakeupAmerica" emphasized that the Islamic background of the shooter played a strong role within the shooting. Others pushed back against such claims with tweets such as "THIS

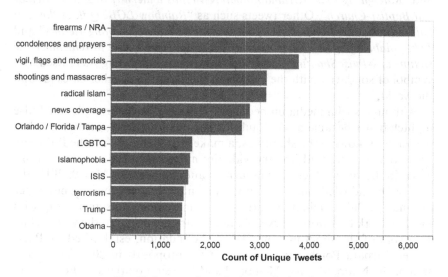

ILLUSTRATION 12.1   Top Twitter topics relating to Pulse shooting.

ATTACK IS NOT AN EXCUSE TO JUSTIFY YOUR ISLAMOPHO-
BIA", and "For hate mongers, any such tragedy is an opportunity to spread
#Islamophobia".

There were many tweets debating the language used by different politi-
cians in the United States about the shooting. For example, many tweets di-
rectly quoted former President Trump's speech in support of the anti-radical
Islam discussions, *"Radical Islam is anti woman, anti gay and anti American -
@realDonaldTrump on #terrorism"*. Others discussed were of former US
Senator John McCain's criticism of Obama's support for gun control: "John
McCain says Obama is 'directly responsible' for #OrlandoShooting, while
John McCain still accepts no responsibility".

A large part of the Twitter conversation expressed condolences to the victims. Tweets mentioned the victims without reference to their membership
of the LGBTQ community: *"Thoughts and prayers for all those affected by
the #OrlandoShooting"*. Others conversations such as *"#OrlandoShooting
was an attack on LGBTQ folks. Saying anything else means, as usual, gays/
lesbians/trans are not worth covering"*. and *"It's tragic that it takes 50+ LG-
BTQ people being massacred for some people to act like they care about the
LGBTQ community"*, raised awareness in solidarity and sympathy to the
members of the LGBTQ community.

Social media was also a space for organizing symbolic and spontaneous vigils and memorials for the victims in Orlando and across the United
States. Conversations of vigils, flags, and memorials highlight how users
welcomed mourners to specific places in a coordinated attempt to remember the victims of the shooting. Examples include: *"Candlelight vigil at Memorial Park for the victims of #OrlandoShooting"*, *"Candlelight vigil is being
held at Lake Eola in Downtown Orlando to mourn after the #PulseShooting"*,
and *"Raleigh vigil for #OrlandoShooting victims underway at Pullen Memorial Baptist Church"*. Other tweets such as *"Rainbow LGBTQ flags flown at
half mast in west London in solidarity with #OrlandoShooting victims"* and
*"The rainbow flag is at half-mast at @Sydney_Uni today, to remember the
victims of #PulseShooting"* described how the rainbow flag was raised as a
symbol of solidarity with the victims of the shooting in other cities around
the world.

Although social media immediately became a place for vigils, the Pulse
nightclub site became a local hub for various long-term, memorials. Immediately following the shootings, a makeshift memorial on the Pulse site
arose that featured ad hoc individually made memorials to the victims.
This site became a focus for mourners and friends. On May 9, 2018, the
City of Orlando dedicated a temporary memorial, consisting of a new fence
around the Pulse nightclub featuring a commemorative screen-wrap with
local artwork that would serve as a temporary memorial to the victims and
survivors of the shooting. OnePULSE Foundation, established by Pulse
owner Barbara Poma, made a request for proposals in 2019 for the design of the National Pulse Memorial and Museum on the site. Forty-eight

national and international design firms submitted proposals, which were later narrowed down to six finalists and the eventual selection of Coldefy & Associés Architectes Urbanistes with RDAI. Construction is expected to begin in 2021.

The relationship between the urban form and social media engagement is an important factor in creating a contemporary memorial. In memory studies, Cherasia compares two AIDS memorials,[9] one in material form and one in the form of a digital archive. The *NAMES Project AIDS Memorial Quilt*, a collection of 48,000 individual quilts displayed in many large public spaces, for example, the entire Washington Mall. The Instagram account *TheAIDSmemorial* had 100,000 followers and 6000 commemorative posts. While comparing the memorials, her research shows that the social media affords more interactivity and reach among users, although the materiality, metaphoric origins, and scope of the Quilt cannot be rendered on digital platforms, representing a loss in affective engagement.

Similar issues of material presence and engagement are discussed by Lowenthal[10] in the context material preservation. He notes contradictions including issues of authenticity, original substance versus form, historic continuity felt by multiple audiences, and separating the past from the present. Other perspectives include preferring fragments to the whole (e.g. the unrestored Parthenon), process to material (e.g. the Ise Shinto temple), written descriptions rather than physical objects (e.g. Chinese tradition for remembering the past). All these options imagine a dynamic and shifting relationship between the materiality of an object and people's engagement.

Cultural memory, a combination of memory, culture, and society in search of meaning[11] and subject to development over time, plays a role in understanding users engagement with online memorials. Mnemonic imagination proposes the recreation of the past as a memory-in-process. As several authors have noted,[12] this is particularly true of social media, a landscape of mass self-communication in constant change and revision. The result is a connective memory open to change over time and reinterpretation by shifting audiences, and a reformulated understanding of the relationship between memory and imagination.

A study of roadside memorials by Klaassens[13] investigates several factors of direct relevance to the Pulse memorials. These locations gain meaning from the event that occurred, and form a focus for parents, friends, and relatives to remember their child and find solace. After some time, more permanent memorials are added to the location, sometimes incorporating religious symbols. One important distinction for the creators of these memorials is the distinction between "good" deaths (after a long life or a heroic act) and "bad" deaths (dying young and by accident).

In *The emotional life of contemporary public memorials*, Erika Doss[14] articulates a more general argument about the contemporary "memorial mania" of public commemoration relative to changing cultural and social practices of mourning, memory, and public feeling. She focuses on temporary memorials of ephemera of flowers, toys, written messages, and cards placed at sites of tragic death, some of them vast in extent (e.g. 200,000 artifacts in front of Buckingham Palace on the death of Princess Diana). The material of the site is cataloged in its entirety, often available in digital form later. She also notes the shift from monuments that commemorate collective public beliefs to memorials that focus on affective engagement[15] with a private vocabulary of grief rather than a common vocabulary of shared values. She argues that "...memorials.... are the physical and visual embodiment of public affect".

To study how the memorials were experienced through social media, a dataset of 1,056 Instagram posts with the hashtag #pulsememorial were collected. The dataset included images, descriptions, and date of posting. The images were manually categorized into several groups based on whether they were associated with either ad hoc, interim, or permanent proposal memorial spaces related to the mass shooting. Two extra groups of other related memorials and digital memorials were identified. The group "Other" refers to images that were about other memorials or vigils from around the world. Several of the images from this group related to pride parades in New York City, or other smaller memorials in Orlando. The "digital" group refers to images that had no relationship to space. In other words, these were purely digital memorials shared by users that were void of urban space. The images and descriptions were qualitatively evaluated for similarities.

Illustration 12.2 shows how the temporal nature of the social media posts about the Pulse memorial follows the development of temporary to permanent memorials over time. Posts related to the ad hoc memorial are prevalent immediately after the shooting up to the point where the nightclub is transformed into the interim memorial. Posts from the interim memorial continue to this day. Spikes in activity mostly correspond to the anniversary of the mass shooting. The images related to the permanent memorial design only show up in 2020 when the winning design was announced. The other two groups have more consistent temporal trends with spikes around the anniversaries of the mass shooting.

This succession of temporal activity for the three ad hoc, interim, and permanent groups, hints at the fact that experiences of visitors are often shaped by the designers and decision-makers who conceive these spaces. While the "other" and "digital" groups show memorialization that mostly corresponds to the symbolic experiences from the LGBTQ communities.

Images associated with the ad hoc group of images were strongly colorful and diverse. The ad hoc memorial consisted of fences around the nightclub that were embellished by visitors with flowers, rainbow flags, and notes. Most images captured a wide view of the site that included the tall and black sign of the Pulse nightclub. A considerable number were close-up images of the items placed by visitors such as colorful stones, different organizations of rainbow flags, a Christmas tree, candles, and small printed rainbow flag

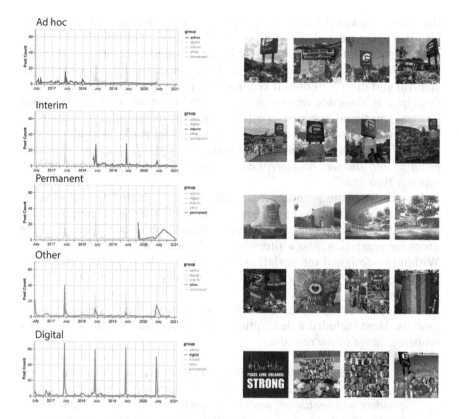

ILLUSTRATION 12.2 Instagram posts of #pulsememorial showing timeline and images.

notes left by visitors. Most of the images from this group were posted before the construction of the interim memorial.

Many descriptions with the ad hoc memorial images were emotional and hopeful and described a visit to the memorial. One of the posts that showed a close-up image of a rainbow set up by the temporary fence with several flowers, is a great example of this hopeful mourning: *"Today I had the opportunity to visit Pulse for the first time since the shooting. To see so many messages of love and unity was inspiring and breathtaking"*.

The interim memorial shifted the types of images posted by visitors. For the construction of the interim memorial, all the colorful items left by visitors were removed. The new construction around the nightclub was organized and consisted of curved walls with printed images related to the LGBTQ community, as well as the victims. The largest printed image in the interim memorial was a typographic poster with the words: "We will not let hate win". A considerable number of images associated with this group were close-ups or selfies taken alongside the large typographic poster. Other images were wider and focused on the curves of the wall. Many of the images showed the base of the nightclub's sign covered in notes from the visitors. There were also a number of selfies taken with the Pulse signage and the notes, although none of the images showed close-ups of the notes. Overall, the new interim memorial and the associated images were not as colorful and diverse as ones taken from the previous ad hoc memorial. The descriptions alongside images of the interim memorial similarly described visitors' emotional experiences after visiting the memorial. For example, a post showing a wide view of the interim memorial read: *"Was able to stop and visit the Pulse memorial today for the first time. […] It was absolutely beautiful and also very heartbreaking to go and see and feel the energy of those who lost their lives"*.

Images from the permanent memorial mostly related to the winning design that is yet to be constructed; therefore, they were computationally rendered and artificial. The predominant colors of the images were green for vegetation, and white, which is the color of the building's facade. Within the designed memorial, and the images, there is little color compared to the ad hoc memorial, or diverse imagery compared to the interim memorial.

The descriptions with these images also contrasted the previous two memorials. Most included a description of the proposed building alongside a rendered image of the new site.

> *The Memorial and Museum will honor the 49 lives that were taken, their families, the 68 injured victims, all the affected survivors, the first responders and healthcare professionals who cared for the victims during the shooting that took place in the Pulse nightclub in Florida on June 12, 2016. The team was required to state how they believe architecture might embody the mandate « We will not let hate win ».We are honoured.*

The "Other" and "Digital" groups of images are colorfully themed. The "Other" group highlights tributes of individuals from around the United States. Almost all images include representations of the rainbow flag buildings and sidewalks painted with rainbow colors, as well as several murals with LGBTQ-related themes. The "Digital" group represents the largest number of images and are the most diverse. A large number included selfies from the LGBTQ community and collaged digital images showing the images of the victims.

The Pulse nightclub is located on an ordinary site along a commercial strip in Orlando. After the shooting there was widespread support for transforming this urban space into a memorial, particularly from the tight local network of LBGTQ advocates many of whom patronized the Pulse nightclub. The wider local community as reflected in the press reported on the politics of this transformation, recording the financial, operational, and strategic issues in each of the iterations of the memorial. This secession of three types of memorials transformed the urban space with different reactions from the local community as reflected on social media.

The creation of a "database" of individual expression of grief and the enhanced importance of the memorial site presents a range of intersections of material form and emotional commentaries. The Pulse nightclub was a well-known location for the LGBTQ community in Orlando and tight-knit network of personal relationships forged over many years. The explosion of social media posts as a result of the killings on June 12 is one measure of its strength. As might be anticipated, the focus of social media was a combination of rage towards politicians and policy-makers, as well as sympathy for the victims and a desire to find a tangible, collective location to express solidarity.

Expressed in terms of urban space, there was the desire to transform an anodyne location on a commercial strip into one of symbolic importance. The LGBTQ communities' colorful symbols within the ad hoc memorial transformed the monotone nature of the site into a place of beauty, hope, and diversity. The interim memorial was more organized, less colorful, but with some traces of the communities' collective identity. Digital memorials, as well as other small vigils were all visually and experientially similar to the representations of the pulse ad hoc memorials with a strong presence of rainbow colors as a symbol of the LGBTQ identity.

The permanent memorial proposal design is white, grand, and iconic. The design is a stark contrast to the lived colorful experiences of the members of the LGBTQ community evident from the Instagram images. The omission of essential references to the LGBTQ community suggests a disconnect between the users of the space and the designers. Urban space in the permanent proposal aligns with Lefebvre's notion of conceived space that is planned and organized by experts that are emotionally, culturally, and physically distant from the intended user's realities.

Notes

1 https://www.washingtonpost.com/news/worldviews/wp/2014/06/18/isis-or-isil-the-debate-over-what-to-call-iraqs-terror-group/.
2 Awan, Imran. (2017). "Cyber-extremism: Isis and the power of social media." *Society, 54*(2), 138–149.
3 https://www.cnn.com/2016/06/13/us/orlando-shooter-omar-mateen/index.html.
4 https://abcnews.go.com/US/multiple-injuries-shooting-orlando-nightclub-police/story?id=39789552.
5 https://www.cnn.com/2016/06/12/us/orlando-shooting-timeline/.
6 https://www.businessinsider.com/james-comey-orlando-shooting-omar-mateen-fbi-2016-6.
7 https://time.com/4371910/orlando-shooting-omar-mateen-facebook/.
8 https://github.com/ddangelov/Top2Vec.
9 Cherasia, Spencer P. (2020). "Affordances, remediation, and digital mourning: A comparative case study of two AIDS memorials." *Memory Studies.* 1750698019894686.
10 Lowenthal, David. (1989). "Material preservation and its alternatives." *Perspecta 25*, 67–77.
11 Assmann, Jan & John Czaplicka. (1995). "Collective memory and cultural identity." *New German Critique, 65*, 125–133.

12 Kaun, Anne & Fredrik Stiernstedt. (2014). "Facebook time: Technological and institutional affordances for media memories." *New Media & Society, 16*(7), 1154–1168.

Castells, Manuel. (2007). "Communication, power and counter-power in the network society." *International Journal of Communication, 1*(1), 29.

13 Mirjam Klaassens, Peter Groote, & Paulus P. P. Huigen. (2009). "Roadside memorials from a geographical perspective." *Mortality, 14*(2), 187–201,

14 Doss, Erika. (2008). *The Emotional Life of Contemporary Public Memorials: Towards a Theory of Temporary Memorials.* Amsterdam: Amsterdam University Press.

15 For a more focused discussion of monument versus memorial, see also Jackson, John Brinckerhoff. (1980). *The Necessity for Ruins, and Other Topics.* Amherst, MA: University of Massachusetts Press.

13

CONCLUSION

The case studies presented in this book explore a range of methods for the analysis of social media data and its influence on the contemporary city. The widespread use of social media has led to shifts in the network of relationships among the local community, social media, and urban space. This translation of social and technological actors has taken a number of forms.

Social media has shaped the way urban space is occupied and used. The city as a stable and unchanging (or at least slowly changing) form with a clear spatial hierarchy is less critical to the way people navigate and inhabit space (i.e. shopping, dining, recreation, and working). Unfamiliar spaces are also now accessible in new ways that promote exploration and host new types of functions. Urban spaces become undiscovered opportunities or refuges to avoid rules of the formal economy. Specifically, underground restaurants create public space in private residences and in abandoned or marginally located buildings, usually under the radar of officials. Mobile food trucks can create new public spaces in empty urban lots, configuring themselves as temporary public squares for communities to gather. Banksy's installations are at their core about the shifting readings of urban space, turning back alleys and previously bland walls into ad hoc museums. This is an example of a private local community of the artists activating marginal urban space with a broad public through social media.

Social media, however, can still enable social activity in well-known and familiar urban spaces, especially historically important public spaces. The emergence of new possibilities for locations within urban space does not signify a complete translation of urban space by social media. For example, public spaces at critical nodes of the city spurred on by an active local community were significant to protesters in Hong Kong, the Women's March in Washington, and the second phase of the Black Lives Matters protests in Charlotte. The popularity of central urban spaces in these protests is based

DOI: 10.4324/9781003026068-17

on a shared understanding of their symbolic and historical position in the community. The proximity of urban space to government buildings and officials, its memory as a site of previous public protest, and its familiarity as a gathering point of community activity all contribute to its repeated use in the central city.

Social media in combination with a local community often supports the temporary re-creation of urban space. Settings are temporary and episodic allowing multiple programmatic uses to inhabit urban space in the course of a day. While not a new phenomenon, in the contemporary city ephemeral urban space is often supported by social media rather than by mass media or liturgical calendars. The spatial configuration of mobile food trucks in empty or marginal urban space is made feasible by the use of Twitter and Roaming Hunger to announce events. The non-profit Burning Man Project is exclusively present on Facebook for 50 weeks followed by two weeks on-site at Black Rock City. Occupation of the site during the festival generates a vast amount of social media data for the next annual social media phase. The switch of digital presence to short-term physical presence can be read as an annual translation that serves to give the Burning Man community energy and renewal.

Social media can enable greater mobility and access throughout the city. Mobile food trucks move to new locations regularly, and social media aids in announcing these changes to customers. In his New York residency, Banksy created daily events in new locations using social media to announce each new piece of artwork, attracting large numbers of visitors instantly. In Hong Kong, the protest that lasted for an entire year changed locations many times to reflect the political climate, tactical measures to avoid the police, and new strategic goals. In the Black Lives Matter protests new locations were communicated to the other protesters through a wide range of social media ranging from open channels such as Facebook and Twitter to encrypted media such as WhatsApp, to the surreptitious use of dating and gaming platforms. Social media afforded the protesters the ability to continue their march regardless of police surveillance.

Social media, as well as many other interfaces are heavily infused with spatial metaphors, a method of providing a structure to abstract ideas and structures. This idea can be seen in William Gibson's use of the "cyberspace" metaphor in his novel *Neuromancer*.[1] More broadly, George Lakoff[2] and Mark Johnson believe our "…ordinary conceptual system … is fundamentally metaphorical in nature…" many of which are spatial. This translation and interaction of spatial metaphors with the structure of social media is an important aspect of the online and on-site memorials for the Pulse nightclub shootings. The range of urban spaces included ad hoc personal remembrances as a roadside memorial, the temporary memorial as organized billboards of images, and the permanent museum as a large-scale architectural sculpture. Each implies a different role for urban space as a keeper of memory. In Iran, the translation of selling goods using social media is an important way for street vendors to navigate their uncertain economic, political, and social status that may be scrutinized in urban space. On one level this metaphor is connected to the specific market locations in a city. On another level, the metaphor of street vendors can be suggested by the wording of online text that recalls a memory of spoken phrasing typical of that used by vendors. Existing urban structures and actions provide a metaphorical structure for these often hybrid online and urban ventures.

Social media activity can be provoked by users' presence in urban space. Images from the Women's March in Washington is, for example, contain political posters carried by protesters within the urban space of the protest. These images were posted and reposted for months after the event carrying forward the importance of both the location and the message. Mobile food trucks also demonstrate an increase in Twitter activity from chefs and patrons about food offerings when the trucks are open. During the Black Lives Matter protests, the social media activity was influenced by participation at the site of the protests, translating activities in urban space into social media. This effect was true for local users, as opposed to non-local users who would not have been at the site. The case of Burning Man presents the opposite case where the urban space and occupation of Black Rock City is experienced not as a location that spurs social media conversation but rather as a location to escape the digital grip.

Social media supported by an active local community enabled a virtual recreation of urban space and events during the cancellation of Burning Man 2020 due to the Coronavirus pandemic. In addition to providing an alternative way to participate in building the Burning Man online community, social media presented a unique opportunity to study how the physical city and in-person festival transformed into the virtual "Multiverse". Despite their differences in style and underlying digital modelling methods, these virtual universes uniformly recreate urban form (or a variant) as the structure of the city. Black Rock City as an urban form is important both for the historical 32-year memory but more importantly as a recognizable pattern of events and activity. Black Rock City has translated into a "memory city" that gives structure to the episodic and virtual experience.

Social Media can be used by local communities for real-time streaming of events on social media. For example, Facebook Live is initially unedited and helps to build the audience for protests that spring from spontaneous events. The translation of the role of mass media into social media led both to the spread of the event and a robust online discussion involving large numbers of users. The postings by the shooter in Christchurch and by the wife of the victim in Charlotte directed enormous audiences to these events. In Charlotte, the posts mobilized protesters for three days. In the case of Christchurch, the shooting video was quickly taken down by Facebook, but served to provoke a strong reaction by New Zealanders supporting the Mosques. The Mosques became an important focus for the online sentiments of solidarity and sympathy combining their everyday function as a gathering spot with a focus on social media posting.

Social media also functions for local communities as a strategic tool in organizing in-person protests both in Hong Kong and the Black Lives Matter protests. In both cases there was the need to mobilize support using social media for protest in public space combined with the necessity to avoid interference from the police intent on limiting the protests. This led to the tactical use of social media for different communications. Facebook and Twitter were very successful at recruiting the general public for events. Other non-public social media platforms such as encrypted text messaging apps through WhatsApp were more useful for organizers planning the protests. In the case of Hong Kong, where the police were aggressive about pursuing the protesters, organizers turned to a wide variety of social media used for alternative purposes such as gaming, dating, and Airdrop file-sharing. The types of social media used shifted to keep ahead of police surveillance.

Social media can employ space as a stage for expressing political and social views. For Burning Man, Black Rock City is a remote location suitable for a gathering that is disconnected from the urban grid. The desert is transformed temporarily into a large-scale event that is streamed and photographed across social media. On the other hand, the Women's March at the well-known National Mall created an ongoing record of Instagram activity. Similar to other political protests, urban space can serve as an important stage, easily understood even by those who are not present at every event. In Christchurch, the mosques became important locations for the expression of solidarity and healing by the local community. In these scenarios, social media provides greater access to a wider public to express their positions and opinions.

Social media can support the formation of local and nonlocal networks of users who reveal distinctive topical and emotional emphasis about an event. During the Black Lives Matters protest in Charlotte social media users with proximity to the event have more pronounced sentiments about grief and sadness, as well as stronger positions about the development of memorials than non-local users. Social media users' ability to be near the event in their community creates a particular social media discourse that emphasizes the importance of place.

Extending Social Media Insight

This wide variety of insights made possible by the use of sophisticated analytic methods opens important new ways to understand the contemporary city. However, the reliance on custom coding for each investigation that requires computational literacy limits the range and audience of a bespoke approach. More importantly, computational approaches to create actionable knowledge from big data are complex and multifaceted. Even with the large number of methods used in this book such as NLP, ML, and social network analysis, there still exists a need for human creativity and knowledge creation to make sense of the findings.

One solution is the use of visual analytic systems which creates a symbiosis of the computational power of computers with the pattern and sense-making capacity of human users. Visual analytic systems include the ability to process high-dimensional data sets, present information visually, and encourage user interaction and exploration.[3] The use of visual analytic principles is uniquely suited, for example, to address the issues of inflexibility of data systems that led to the need for improved urban planning support systems.

Insight and knowledge are key aspects of visual analytic systems. There is considerable research on ways human computation systems are most effective, for example, the knowledge generation paradigm research of Daniel Keim.[4] Such systems are particularly useful and appropriate approaches for design. Keim articulates three levels of reasoning or "loops" that form a framework for human computation. An important feature of these "loops" is that they represent not steps in a linear sequence, but rather three activity loops with distinct goals and processes that are engaged iteratively.

The exploration loop is defined as an interaction between the user and the system using feedback to guide the activities of the user. This loop works most effectively when the user has a specific goal and engages a flexible structure for the exploration that can lead to new insights and analytic goals. The verification loop is the creation of a hypothesis or a set of assumptions about relationships between data or categories of data which is testable within the world of the visual analytic system. Insight can result from this interaction between assumptions and findings, which can in turn lead to knowledge. The knowledge generation loop tests and extends the interpretative power of expert users' existing knowledge. This can provide evidence for existing knowledge or lead to new insight and knowledge.

Most systems currently in use that aim to understand the contemporary city have not yet incorporated either social media (which is too heterogeneous for existing systems) or visual analytic interfaces and knowledge systems. Examples that incorporate visual analytic systems include work by Chang et al. on an alternative to Geographic Information System (GIS)[5] to provide an interactive interface for US Census data exploration in combination with a city's urban form characteristics. An example of research toward the development of an exploratory interface for social media data is the Urban Activity Explorer[6] that allows a user to observe the locations of human activity in urban space. To demonstrate the feasibility of this approach, the Urban Activity Explorer used approximately 1 million geolocated tweets from the Twitter Firehose for September and October 2015 for the Los Angeles metropolitan area. Geolocated tweets, specifically, include content, location, and time of activity among other metadata. However, the unstructured nature of the data makes it very difficult for the user to benefit from the information without a close and, for large amounts of data, arduous reading of individual tweets. In contrast to a top-down system such as

GIS with fixed and predetermined capacities, evaluative criteria, and outcomes, Urban Activity Explorer is designed to allow users to explore disparate forms and patterns of information. This exploratory approach allows participants to gather a variety of insights about a single topic by using varied, interrelated types of information. This interface is designed for the exploration of the spatial, temporal, and topical aspects of social media data as a proxy for human activity (Illustration 13.1).

There are six highly interlinked views within the application interface that reveal human activity. Illustration 13.1 shows an overview of the interface that consists of two map views and four information views. It has six main views: (A) tweet density view, (B) tweet flow view, (C) word cloud view, (D) tweet timeline view, (E) flow length/time view, and (F) tweet language/topic view. The map views (Illustration 13.1A and B) are the primary components and the main points of interaction for the user. The information views (Illustration 13.1 C, D, E, and F) contain information about the two primary views, and they can affect the state of the two map views and are also affected by them. Interaction with each of the views changes the state of other views to allow the users to explore different spatial and temporal aspects of the dataset.

Tweet density view shows the geographic concentration of geolocated tweets for a certain area. The user can zoom in/out and pan over different areas of this map which dynamically updates to hotspots in response to the scale. The user can also view an animation of densities in every hour, or use a time slider to explore the temporal nature of activities during that day through viewing activities by hour.

Word cloud view shows the most frequent words within the user's map view. The view is dynamically calculated based on the current map extent of the tweet density view. Any change in the tweet density view ranging from zooming, panning, time selection, or content selection will update the word

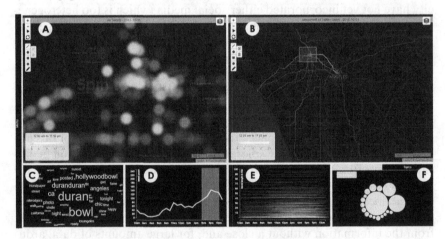

ILLUSTRATION 13.1 The Urban Space Explorer interface.

cloud, allowing the user to dynamically explore the content of tweets from different scales and times.

Tweet timeline view shows a timeline of tweet activity per hour for a day. The timeline responds to the scale on which the user is observing in the tweet density view. The user is able to select a portion of the timeline graph to observe the tweet density for that specific time span and a set of pre-calculated topics that are derived from the tweets using the LDA algorithm.[7] The user can select each topic and view a heatmap, timeline, and word cloud corresponding to that specific topic.

Tweet flow view shows the movement of Twitter users during a certain day. The paths of individual users mapped to street networks reflect human flows within cities using Djistra's shortest path algorithm.[8] Similar to the heatmap view, zooming in and out will update the level of details of the tweet flows. At larger scales, regional travel is more visible and vice versa. The users can select regions and view the flow of people to that region (drawn region as destination) or flow of people out of that region (drawn region as origin). Users can view an animation of flow for the whole day and also use a time slider to explore flows at consecutive time periods during the day.

Flow Length/Time view shows individual movement paths coded by length and time. Each individual line represents one user's travel path. The x-axis of this graph codes the start and end time of travel. The y-axis shows the total length of travel of each line. For example, the lower the line on the graph, the shorter the travel path.

Tweet language view shows a distribution of people who tweet in different languages during the day. The language of tweets is used as a proxy for demographic information. Users can select a certain language and view the distribution of users who tweet in that language (e.g., English vs. Spanish). The size of the circles denote the number of tweets in a certain language.

The topic view is located alongside the language view. This view shows the results of topic modelling of the whole tweet corpus using LDA. The user can explore the topic lists and then view the tweet density, timeline, and word cloud related to that topic.

Urban Activity Explorer offers unique advantages to planners and urbanists dedicated to improving the city. For instance, they can use it to develop a funding strategy for open spaces and public parks by evaluating the users' sentiments about the quality and use of these spaces. Resources for public art can be allocated to areas identified as having cultural diversity and interested residents. Likewise, surveying tweets across main streets for signs of struggling business activity can help determine priority areas in need of reinvestment. Transportation planners could identify areas with an established pedestrian realm and potential market for introducing bike-share systems or locating bike paths. Or they could study areas with frequent traffic accidents. At a larger geographic scale, social media activity showing patterns of migration across time among different cultural groups can support new housing development.

Urban Activity Explorer is a bottom-up tool, using heterogeneous data collected by social media. It affords users the ability to discover urban issues and solutions that may not have been considered previously and presents two important advantages over top-down systems: understanding emergent issues that can easily be missed by goals and methods decided in advance, and the ability to obtain data already streaming in abundance that can allow the system to respond quickly.

Social media data is unique in that it presents temporal, spatial, and topical information together. This affords the opportunity to discover linkages and meanings that are often obscured by the rigid isolation of data on incompatible layers. Social media data are messy, but they are intrinsically interconnected. Urban Activity Explorer includes advanced data analytic methods not previously accessible in a system that is easy to use and does not require advanced technical knowledge by the user. Moreover, risks associated with social media including the privacy of users' data and low quality and accuracy of information are minimized by the aggregation of publicly available data.

Visual analytics offers a rich field of scholarship for understanding how human users and computational power can work together in a complementary system. The rapid development of knowledge discovery noted above within visual analytics provides a theoretical overview and practical methods for providing the kind of support that will allow the integration of meaningful computation to the complex issues facing the contemporary city.

Despite the advantages offered by this system, there are currently no widely available visual analytic systems that can deal with social media data. Several factors inhibit the development of such systems.

First, it is a major challenge for the field to build robust and scalable visual analytic interfaces capable of dealing with the enormous volume of social media data. Proof of concept systems such as Urban Activity Explorer deal with only a tiny fraction of the daily volume of social media and focus on text data rather than image data. It may require a multi-year publicly funded research effort to fully develop such an interface.

A second factor is the hesitancy of social media companies to share the vast amounts of data collected. There is no requirement to share this potential data with researchers. Data are a source of huge profits for social media companies, and they are hesitant to share it. The only social media company that currently allows limited access for research is Twitter. Data have been sold to the highest bidder[9] (for example to Cambridge Analytics during the US presidential election in 2016), but data are generally not available to academic researchers, the press, the public, or regulators. This is a major hurdle in the ability to effectively understand and use the data for research and planning.

A third issue is the privacy of social media users. It is certainly true that social media companies are interested in using the information to generate revenue through targeted advertising and the sale of data. More widespread access to social media data would raise issues that would require protecting individual identity and anonymizing data. A body of research in medical records privacy[10] might provide a basis for resolving this issue.

Social Media and the Contemporary City

The invention of the smartphone in 2006 had a profound impact on social media platforms. Their ability to input and process information that was previously only possible on a desktop computer led to the explosion of users and content on social media platforms such as Twitter (353 million users), Facebook (2.7 billion), and Instagram (1.22 billion).[11] And unlike desktop computers, smartphones go with users everywhere; out on a shopping trip, out to dinner, out to a concert, out in the car, out to work. Social media has become overwhelmingly *mobile* social media and has become implicitly spatial. Fully 92% of Facebook's revenue in 2018 was from mobile devices.[12] Users engage on social media throughout the day, creating a rich trove of data about what they are thinking, where they are, and what they are seeing. Social media has become closely intertwined with the city.

Social media enables all users to contribute directly through interactive interfaces. In systems analysis, mass media such as television and newspapers are characterized as "one to many", while social media is the archetypal example of "many to many". Unlike mass media, control of social media content is not centralized as it is in a book, a television station, or a newspaper. The form of media is also malleable in social media, allowing for extended narrative, simple lists, or uncaptioned photographs. This results in a wide variety of content and commentary including those that may be marginal either by taste, interest, or politics. Marmot farmers, Tibetan stamp collectors, and conspiracy theorists can all find a group on social media. The results of these online communities can be supportive, humorous, or threatening. The lack of an authoritative form is liberating but it comes at the cost of a single or at least curated vision.

Social media is subject to wide variation in content and participation over time. Online communities are not permanent. They can expand and contract over time or with activity in local communities or urban spaces. And they can shift to new social media platforms for strategic or tactical reasons. Events have always been part of urban space, but typically on a yearly, monthly, or weekly basis. Social media has accelerated the creation of events, the scale of events, and how quickly they shift locations. Political protests can now adapt to police tactics minute by minute using real-time information. Events can be at a micro-scale involving a traffic accident or food truck, or at the scale of a simultaneous demonstration in 400 cities. Considerable effort has been made to understand events using the topics and themes of social media data within an online community, which involves tracing their instance, growth, and duration.

This book demonstrates that the formation of online communities through social media has not replaced urban space or local communities.

Instead, they operate in urban space and through the involvement of local communities formed by personal contact and private communications. In the contemporary city, urban space, local community, and social media become interlinked and dependent on each other in new ways. They do not become less relevant but rather they translate into new roles or relationships with other actors. A new network of relationships is being reconstructed that requires an adjusted model and radically expanded methods to understand. And there will inevitably be advantages and dangers that are not yet entirely clear.

Notes

1 Gibson, William. (2015). *Neuromancer* (Vol. 1). Aleph.
2 Lakoff, George & Mark Johnson. (2008). *Metaphors We Live By.* Chicago, IL: University of Chicago Press.
3 Keim, Daniel et al. (2008). "Visual analytics: Definition, process, and challenges." In Andreas Kerren, John T. Stasko, Jean-Daniel Fekete & Chris North (Eds.), *Information Visualization* (4950, pp. 154–175). Berlin: Heidelberg, Springer.
4 Sacha, Dominik et al. (2014). "Knowledge generation model for visual analytics." *IEEE Transactions on Visualization and Computer Graphics, 20*(12), 1604–1613.
5 Chang, Remco, Ginette Wessel, Robert Kosara, Eric Sauda, & William Ribarsky. (2007). "Legible cities: Focus-dependent multi-resolution visualization of urban relationships." *IEEE Transactions on Visualization and Computer Graphics, 13*(6), 1169–1175.
6 Karduni, Alireza et al. (2017). "Urban activity explorer: Visual analytics and planning support systems." *International Conference on Computers in Urban Planning and Urban Management.* Cham: Springer.
7 Blei, David M., Andrew Y. Ng, & Michael I. Jordan. (2003). "Latent dirichlet allocation." *The Journal of Machine Learning Research, 3*, 993–1022.
8 Dijkstra, Edsger W. (1959). "A note on two problems in connexion with graphs." *Numerische Mathematik, 1*(1), 269–271.
9 Ward, Ken. (2018). "Social networks, the 2016 US presidential election, and Kantian ethics: Applying the categorical imperative to Cambridge Analytica's behavioral microtargeting." *Journal of Media Ethics, 33*(3), 133–148.
10 Szarvas, György, Richárd Farkas, & Róbert Busa-Fekete. (2007). "State-of-the-art anonymization of medical records using an iterative machine learning framework." *Journal of the American Medical Informatics Association, 14*(5), 574–580.
11 https://www.statista.com/statistics/272014/global-social-networks-ranked-by-number-of-users/.
12 https://www.statista.com/statistics/271258/facebooks-advertising-revenue-worldwide/#:~:text=In%202018%2C%20Facebook's%20mobile%20advertising,billion%20U.S.%20dollars%20in%202018.

INDEX

Note: *Italic* page numbers refer to figures.

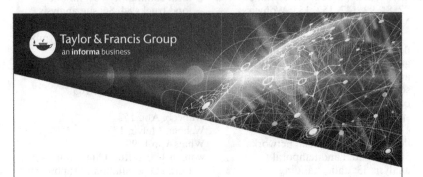

Printed in the United States
by Baker & Taylor Publisher Services